Question Bank for Agricultural Competitions

(J.R.F., S.R.F., N.E.T., C.E.T. *and* Ph.D.)

THE AUTHORS

Mr. Rakesh Kumar Sharma is Assistant Seed Certification Officer in CGSSCA, Raipur. He has also worked as Senior Surveillance Inspector and SRF in Indira Gandhi Krishi Vishwavidyalaya, Raipur. He received his B.Sc (Ag.) in 2006 and M.Sc (Ag.) degree in 2008 in the field of Agronomy from Indira Gandhi Krishi Vishwavidyalaya, Raipur (C.G.). He has qualified JRF (Agronomy) during 2006 and Ph.D. Exams of Banaras Hindu University & GBPUAT during 2009-10 and NET examination during 2010. He has also qualified National Eligibility Test (ASRB, ICAR). He has published books "Guide for Agricultural Entrance Examinations" and "Agriculture At A Glance", many research papers and technical articles. He has lot of experience of teaching of Under-graduate programme.

Dr. H.P. Agrawal is Principal Scientist in Thakur Chhedilal Barrister College of Agriculture and Research Station, IGKV, Bilaspur. He started his carrier as subject matter specialist in Krishi Vigyan Kendra after that he has worked as district agronomist in IERP project. He also holds the post of Assistant Director of Agriculture and manages the farm. He received his B.Sc. (Ag) and M.Sc. (Agronomy) degree from Jawahar Lal Nehru Krishi Vishwavidyalaya, Jabalpur, and Ph.D. (Life Sciences) from Guru Ghasidas University, Bilaspur. He has also received L.LB from Guru Ghasidas University, Bilaspur and done M.B.A. (Ag.) from Sam Higinbottom Institute of Agriculture Technology and Sciences, Deemed University, Allahabad. He has vast experience of teaching of undergraduate and Ph.D. programme and research and extension activities. He has published many research papers and technical bulletins. He has received many awards in the field of agriculture.

Dr. Geet Sharma received his B.Sc. (Ag.) degree from Jawahar Lal Nehru Krishi Vishwavidyalaya, Jabalpur, M.Sc.(Ag.) and Ph.D (Agronomy) from Indira Gandhi Krishi Vishwavidyalaya, Raipur. He has started his carrier as technical assistant in Krishi Vigyan Kendra and after that in AICRP on Agroforestry. Presently he is working as scientist (Agronomy) at Thakur Chhedilal Barrister College of Agriculture and Research Station, IGKV, Bilaspur. He has vast experience of teaching of Undergraduate and research and extension activities. He has published many research papers in reputed journals and technical bulletins.

Mr. Naval Kishor Singh Narwariya is Rural Agriculture Extension Officer at Department of Agriculture, Kawardha (C.G.). He received his B.Sc (Ag.) in 2006 and M.Sc (Ag.) degree in 2008 in the field of Agronomy from IGKV, Raipur (C.G.). He has published many research papers and technical articles in many journals. He has lot of experience in teaching and extension activity in the field of his specialization.

Mr. Ajay Kumar Tripathi is Assistant Seed Certification Officer in CGSSCA, Raipur. He also worked as Rural Agriculture Extension Officer at Department of Agriculture, Kawardha (C.G.). He received his B.Sc (Ag.) in 2006 from IGKV, Raipur (C.G.) and M.Sc (Ag.) degree in 2008 in the field of Plant Breeding from BCKVV (W.B.). He has published many research papers and technical articles in many journals. He has well known in teaching and extension activity in the field of his specialization.

Question Bank for Agricultural Competitions

(J.R.F., S.R.F., N.E.T., C.E.T. *and* Ph.D.)

Useful for:

- ☆ *JRF, SRF and NET Examinations of ICAR*
- ☆ *M.Sc. and Ph.D. Examinations of IGKV, BHU, GBPUAT and various Indian Agricultural Universities*
- ☆ *ADA and Bank AO Examinations*

Author
R.K. Sharma

Co-Authors
H. P. Agrawal, G. Sharma, N. K. S. Narvaria and A. Tripathi

2021

Astral Cracker

A Division of

Astral International (P) Ltd

New Delhi 110 002

First Impression, 2013
Reprinted, 2021

ISBN 978-81-7035-791-9 (PB)

Publisher's note:
Every possible effort has been made to ensure that the information contained in this book is accurate at the time of going to press, and the publisher and author cannot accept responsibility for any errors or omissions, however caused. No responsibility for loss or damage occasioned to any person acting, or refraining from action, as a result of the material in this publication can be accepted by the editor, the publisher or the author. The Publisher is not associated with any product or vendor mentioned in the book. The contents of this work are intended to further general scientific research, understanding and discussion only. Readers should consult with a specialist where appropriate.
Every effort has been made to trace the owners of copyright material used in this book, if any. The author and the publisher will be grateful for any omission brought to their notice for acknowledgement in the future editions of the book.

Published by : **Astral Cracker**
A Division of
Astral International Pvt. Ltd.
– ISO 9001:2015 Certified Company –
4736/23, Ansari Road, Darya Ganj
New Delhi-110 002
Ph. 011-43549197, 23278134
E-mail: info@astralint.com
Website: www.astralint.com

Printed at : **Replika Press Pvt Ltd**

PRINTED IN INDIA

Preface

Agriculture plays an important role to cover primary aims of mankind like food, fibre, fuel etc by optimum use of terrestrial resources. The Indian Council of Agricultural Research (ICAR), New Delhi is conducting All India Competitive Examination every year for awarding JRF, SRF, NET, ARS etc. in addition to that Common Entrance Test for competitive examinations for post-graduate degree, Doctor of Philosophy courses of various State Agricultural Universities, RAEO, ADO, SEDO, ADA of State Boards, Administrative and Banking AO Services are being conducted for selection of suitable candidates. All theses examinations are mostly objective based and students always look for study material that is ready to use and easy to grasp.

It is facing that there is hardly any book available in the market on agriculture which completely satisfy the requirement of the students particularly for those who are preparing for competitive examinations.

Keeping in view, the prevailing situation, it was deeply felt to publish such a book which could serve the basic and innovative knowledge of various fields of agriculture.

The present book *"Question Bank For Agricultural Competitions"* has been prepared in most simple, clean and appropriate manner which covers all courses of competitive made with the students, teachers and scientists. We hope that the book will fulfill your need on agriculture.

The cooperation and encouragement extended by our family members during the completion of this book is highly appreciable.

We are heartily thanks to all our friends who have taken sincere efforts to do creative works like this.

We humbly welcome the valuable suggestions from reader for further improvement of this book.

Rakesh Sharma

Contents

Chapter 1

Current Agricultural Trends

Quest. Total Geographical Area (TGA) of India

Ans. **329 Mha**

Quest. Net cultivated/sown area of India.

Ans. **143 Mha**

Quest. Net Irrigated area in India

Ans. **56.3 Mha**

Quest. Per Capita Agril. land availability

Ans. **0.16 ha**

Quest. Cropping Intensity of India

Ans. **136%**

Quest. Total degraded land of India

Ans. **174 Mha**

Quest. Area threatened by land degradation

Ans. **50% of TGA**

Quest. Area potential for biological production

Ans. **256 Mha**

Quest. % geographical area of India used for agricultural activity

Ans. **43%**

Quest. India's position on world's total population

Ans. **2nd (1st - China)**

Quest. India's support on forest area over world

Ans. **1.5%**

Quest. As per National Forest Policy, the forest cover of the country should be on

Ans. **33%**

Quest. India's support on world's total livestock

Ans. **15%**

Quest. India's support on world's TGA

Ans. **2%**

Quest. India's support on world's total water resources

Ans. **4%**

Quest. India's position on world's total agricultural area (2009-10)

Ans. **7th**

Quest. India's position on world's total arable land (2009-10)

Ans. **2nd (159 Mha)**

Quest. India's position on world's total irrigated area (2009-10)

Ans. **1st**

Quest. India's position in world's total cereal production (2009-10)

Ans. **3rd (China > USA > India)**

Quest. India's position on world's total rice production (2009-10)

Ans. **2nd (1st - China)**

Quest. India's position on world's total wheat production (2009-10)

Ans. **2nd (1st - China)**

Quest. India's position on world's total coarse grains production (2009-10)

Ans. **4th**

Quest. India's position on world's pulse production (2009-10)

Ans. **1st**

Quest. India's position on world's total oilseed production (2009-10)

Ans. **2nd (1st - China)**

Quest. India's position on world's total cotton production (2009-10)

Ans. **2nd (1st - China)**

Quest. India's position on world's sugarcane production (2009-10)

Ans. **2nd (1st - Brazil)**

Quest. India's position on world's total fruit area and production (2009-10)

Ans. **2nd (1st - China)**

Quest. Rank of India on world's total vegetable production (2009-10)

Ans. **2nd (1st - China)**

Quest. India's rank on world's total milk production (2009-10)

Ans. **1st (108.5 Mt., 2009)**

Quest. India's rank on world's total egg production (2009-10)

Ans. **3rd (55 Billion)**

Quest. India's position on livestock population in world (2009-10)

Ans. **1st**

Quest. Rank of India for transgenic plant in world (2009-10)

Ans. **4th (1st-USA)**

Quest. India's rank on world's total fertilizer consumption (2009-10)

Ans. **3rd**

Quest. India gets 1st position in world's agriculture on

Ans. **Production of Pulse Jute, Buffalo population, milk, irrigated area**

Quest. India gets 2nd position in world agriculture on

Ans. **Rice, Wheat, Groundnut, Sugarcane, Cotton, Fruits and Vegetables, Potato, Onion, Arable land, total population, Cattle and Goats, Tractor in-use**

Quest. India gets 3rd position on world's agriculture on

Ans. **Production of Total cereals, Rapseed, Tea, Tobacco leaves, Sheep, Egg**

Quest. Irrigation potential of India (March, 2010)

Ans. **108.2 Mha**

Quest. State having maximum area under irrigation

Ans. **Punjab**

Quest. Major source of irrigation in India

Ans. **Canal**

Quest. State having highest Net Irrigation Potential in India under canal and tubewell

Ans. **UP**

Quest. Area under Micro irrigation system in India (2008-09)

Ans. **3.96 Mha**

Quest. Area under Drip irrigation in India (2008-09)

Ans. **1.42 Mha**

Quest. State having highest area under Drip irrigation in India

Ans. **Maharashtra**

Quest. Area under Sprinkler irrigation in India (2008-09)

Ans. **2.54 Mha**

Quest. State having highest area under Sprinkler irrigation in India

Ans. **Haryana**

Quest. % National water use efficiency

Ans. **40%**

Quest. % growth rate in production of Indian agriculture

Ans. **5.7%**

Quest. Which crop having highest acreage in the world?

Ans. **Wheat > Rice > Maize**

Quest. Which crop having highest acreage in India?

Ans. **Rice > Wheat > Maize > Sorghum**

Quest. Leading crops in total acreage and production in India

Ans. **Rice > Wheat > Maize**

Quest. Crop having the maximum yield potential in the world

Ans. **Maize**

Quest. Crop having the maximum yield potential in India

Ans. **Wheat**

Quest. Crop having highest Net sown Irrigated area

Ans. **Sugarcane**

Quest. Crop having highest Grass cropped Irrigated area

Ans. **Wheat**

Quest. Irrigated area under sugarcane in India (2008-09)

Ans. **4.5 Mha (93.7%)**

Quest. Irrigated area under wheat in India (2008-09)

Ans. **25.5 Mha (91.3%)**

Quest. Irrigated area under rice in India (2008-09)

Ans. **26.5 Mha (58.7%)**

Quest. Production of total food grains in India (2009-10)

Ans. **218.20 Mt.**

Quest. Production of Rice in India (2009-10)

Ans. **89.13 Mt.**

Quest. Leading state in production of Rice in India (2008-09)

Ans. **WB > AP > UP**

Quest. Production of Wheat in India (2009-10)

Ans. **80.71 Mt.**

Quest. Leading state in production of Wheat in India (2008-09)

Ans. **UP > Punjab > Haryana**

Quest. Production of Maize in India (2009-10)

Ans. **19.73 Mt.**

Quest. Leading state in production of Maize in India (2008-09)

Ans. **AP > Karnataka > Rjs**

Quest. Production of total course cereals in India (2009-10)

Ans. **33.77 Mt.**

Quest. Leading state in total course cereals in India (2009-10)

Ans. **Rjs > Karnataka > Maharashtra**

Quest. Production of total pulse in India (2009-10)

Ans. **14.66 Mt.**

Quest. Leading state in production of total pulses in India (2008-09)

Ans. **MP > UP > Rjs**

Quest. Leading state in production of Groundnut in India (2008-09)

Ans. **Gujarat > AP > TN**

Quest. Leading state in production of Rapseed and Mustard in India (2008-09)

Ans. **Rjs > UP > Haryana**

Quest. Leading state in production of Soybean in India (2008-09)

Ans. **MP > Maharashtra > Rjs**

Quest. Leading state in production of Sunflower in India (2008-09)

Ans. **Karnataka > AP > Maharashtra**

Quest. Production of total oilseeds in India (2009-10)

Ans. **24.93 Mt.**

Quest. Leading state in production of total oilseeds in India (2008-09)

Ans. **MP > Rjs > Gujarat**

Quest. Production of Cotton (2009-10)

Ans. **24.22 Mt.**

Quest. Leading state in production of Cotton in India (2008-09)

Ans. **Gujarat > Maharashtra > AP**

Quest. Production of Sugarcane in India (2009-10)

Ans. **292.30 Mt.**

Quest. Leading state in production of Sugarcane in India (2008-09)

Ans. **UP > Maharashtra > TN**

Quest. Production of Jute and Mesta in India (2009-10)

Ans. **11.82 Mt.**

Quest. Leading state in production of Potato in India (2008-09)

Ans. **UP > WB > Punjab**

Quest. Area under Bt-cotton in India (2009)

Ans. **8.4 Mha**

Quest. % Share of crops in Total Horticultural area of India (2009-10)

Ans. **Vegt. (40%) < fruits (30%) < Plantation (15%)**

Quest. Leading fruits in area in India (2009-10)

Ans. **Mango > Citrus > Banana**

Quest. Leading state in fruits crop area in India (2009-10)

Ans. **MH > AP > UP**

Quest. Leading fruits in production in India (2009-10)

Ans. **Banana > Mango > Citrus**

Quest. Leading states in fruits crop production in India (2009-10)

Ans. **AP > MH > TN**

Quest. Leading fruits in productivity in India (2009-10)

Ans. **Papaya > Banana > Grape**

Quest. Leading vegetable in area in India (2009-10)

Ans. **Potato > Tomato > Onion**

Quest. Leading state in vegetable crop area in India (2009-10)

Ans. **WB > UP > Bihar**

Quest. Leading states in vegetable production in India (2009-10)

Ans. **WB > UP > Bihar**

Quest. Leading vegetables in productivity in India (2009-10)

Ans. **Tapioca > Cabbage > Potato**

Quest. Which two crops together constituted 78 % of total foodgrains production in India (2009-10)

Ans. **Rice and Wheat**

Quest. Which two crops are the major contributors to the total pulse production?

Ans. **Gram and Arhar**

Quest. Contribution of Indian agriculture to livelihood

Ans. **65%**

Quest. Share of agriculture and allied sectors to the National GDP (2009-10)

Ans. **14.6%**

Quest. Share of livestock and fisheries to the National GDP (2009-10)

Ans. **4.07%**

Quest. Share of agriculture to National exports

Ans. **10.23% (2008-09)**

Quest. Share of agriculture to National imports

Ans. **2.74% (2008-09)**

Quest. Annual export of high quality basmati rice from India

Ans. **0.5-0.7 Mt.**

Quest. Export of non-basmati rice has been prohibited since

Ans. **15 October 2007**

Quest. Export of wheat has been prohibited since

Ans. **8 October 2007**

Quest. All India avg. fertilizer consumption is

Ans. **128.8 Kg/ha (2008-09)**

Quest. Highest avg. fertilizer consumption is found in

Ans. **Punjab (212Kg/ha)**

Quest. Nutrient consumption ratio of India (2009)

Ans. **6.5 : 2.5 : 1**

Quest. 100% imported fertilizer in India

Ans. **Potash (K_2O)**

Quest. % Govt subsidy provided to farmers on fertilizers prices

Ans. **60-75%**

Quest. Net availability of cereals per capita/day (2009-10)

Ans. **407 gm**

Quest. Net availability of pulses per capita/day (2009-10)

Ans. **37 gm**

Quest. Net availability of Fruits per capita/day (2009-10)

Ans. **120 gm**

Quest. Net availability of Vegetables per capita/day (2009-10)

Ans. **140 gm**

Quest. Net availability of Milk per capita/day (2009-10)

Ans. **263 gm**

Quest. Net availability of Egg per capita/day (2009-10)

Ans. **45 (no)**

Quest. The annual rate of interest received by the farmers on crop loans from govt.

Ans. **5%**

Quest. To be self reliant in agriculture, the necessary growth rate per year is

Ans. **4%**

Quest. ICAR day is celebrated every year on

Ans. **16th July**

Quest. World Food Prize (2009) was awarded to

Ans. **Gebisa Ejeta (Ethiopia) for 1st sorghum hybrid for drought and Striga weed.**

Quest. Indian scientist shared World Food Prize for Miracle Maize

Ans. **Dr. Sruinder K. Vasal**

Quest. Nobel Peace Prize was awarded to Dr. Norman Borlaug in

Ans. **1972**

Quest. National Rural Employment Scheme was started on

Ans. **2006**

Quest. National Agriculture Policy was started on

Ans. **2000**

Quest. National Seed Policy was started on

Ans. **2002**

Quest. National Food Security Mission (NFSM) was launched on

Ans. **Rabi, 2007-08**

Quest. Rashtriya Krishi Vikas Yojana (RKVY) was launched on

Ans. **2007-08**

Quest. National Horticulture Mission (NHM) was launched on

Ans. **2005-06**

Quest. Integrated Scheme on Oilseeds, Pulses, Oil Palm and Maize (ISOPOM) started since

Ans. **1 April, 2004**

Quest. National Mission on Micro Irrigation (NMMI) was launched on

Ans. **June, 2010**

Quest. National Bamboo Mission (NBM) is implemented from

Ans. **2006-07**

Quest. Kishan Credit Card Scheme was launched on

Ans. **1998-99**

Quest. National Agricultural Insurance Scheme was launched on

Ans. **1999-2000**

Quest. Green revolution is mainly related with the crops

Ans. **Wheat and Rice**

Quest. 'Rainbow revolution' refers to

Ans. **Overall development of agril. sectors**

Quest. Yellow revolution is associated with

Ans. **Oilseeds production**

Quest. Operation Flood denotes

Ans. **3 fold increase in milk production in India**

Quest. FCI is specially launched for

Ans. **Rice, Wheat and Course millets**

Quest. Hybrid rice for commercial production was first evolved at

Ans. **China**

Quest. Minimum support price of Paddy (2011-12)

Ans. **1080 Rs/qt**

Quest. Minimum support price of Grade A-Paddy (2011-12)

Ans. **1110 Rs/qt**

Quest. Minimum support price of Wheat (2011-12)

Ans. **1285 Rs/qt**

Quest. Minimum support price of Maize, Hybrid Jowar and Barley (2011-12)

Ans. **980 Rs/qt**

Quest. Minimum support price of Gram and Lentil (2011-12)

Ans. **2800 Rs/qt**

Quest. Minimum support price of Arhar (2011-12)

Ans. **3200 Rs/qt**

Quest. Minimum support price of Moong (2011-12)

Ans. **3500 Rs/qt**

Quest. Minimum support price of Urd (2011-12)

Ans. **3300 Rs/qt**

Quest. Minimum support price of Cotton (2011-12)

Ans. **2800 Rs/qt (F-414/H-777, J34) and 3300 Rs/qt (H-4)**

Quest. Minimum support price of Soybean (2011-12)

Ans. **1650 Rs/qt (Black) and 1690 Rs/qt (Yellow)**

Quest. Minimum support price of Mustard and Sunflower (2011-12)

Ans. **2500 Rs/qt**

Quest. Minimum support price of Safflower (2011-12)

Ans. **1800 Rs/qt**

Quest. Minimum support price of Jute (2011-12)

Ans. **1600 Rs/qt**

Quest. Minimum support price of Sesamum (2011-12)

Ans. **3400 Rs/qt**

Quest. Minimum support price of Groundnut in shell (2011-12)

Ans. **2700 Rs/qt**

Quest. Minimum support price of Sugarcane (2011-12)

Ans. **139.12 Rs/qt**

Quest. First agriculture census in India conducted in

Ans. **1970**

Quest. First livestock census in India conducted in

Ans. **1919**

Quest. First All-India Co-ordinate Research Project (ACRIP) on

Ans. **Maize (1957)**

Quest. First State Agricultural University of India

Ans. **GBPAUT, Pantnagar (1960)**

Quest. First Krishi Vigyan Kendra (KVK) was established at

Ans. **Puducherry (Pondicherry, 1974)**

Quest. Total no. of KVK in India

Ans. **568 (Dec.2009)**

Quest. Union Minister of Agriculture (2010-11)

Ans. **Sharad Pawar**

Quest. New Director-General of ICAR (2010-11)

Ans. **Dr. S. Ayyappan**

Quest. Chairman of Agricultural Scientists' Recruitment Board (ASRB) of ICAR

Ans. **C. D. Mayee**

Quest. Insecticidal Act was passed by the Government of India in

Ans. **1968**

Quest. Pesticides restricted for use in India

Ans. **13**

Quest. No. of Insecticides approved to control household pests

Ans. **39**

Quest. The rice having richness in beta-carotene and also contain vitamin A

Ans. **Golden rice**

Quest. The rice which can alleviate anaemia problem through dietary intake

Ans. **Ferritin rice**

Quest. The genetic modified egg with medicinal values is

Ans. **Golden egg (developed in Australia in 1999)**

Quest. Irritation of eye due to cutting onion is corrected by

Ans. **Super Sweet Onion (developed in UK)**

Quest. 'Indian farming' is a publication from

Ans. **ICAR**

Chapter 2
Agronomy

(I) Basic Principles of Crop Production

Quest. A very broad term encompassing all aspects of crop production, livestock farming, fisheries, forestry etc.

Ans. **Agriculture**

Quest. '*Agriculture*' word is derived from

Ans. ***Latin word* (*agri+culture*)**

Quest. A branch of agricultural science which deals with principles and practices of soil, water and crop management.

Ans. **Agronomy**

Quest. "Agronomy" word is derived from words?

Ans. ***Greek* (*agros+nomos*)**

Quest. Crops which are cultivated on ploughed land?

Ans. **Arable crops**

Quest. An agroforestry practice in which perennial, preferably leguminous trees or shrubs are grown simultaneously with arable crop?

Ans. **Alley crops or *hedge-row intercrops***

Quest. Crops which are grown to supplement the yield of the main crops?

Ans. **Augment Crops**

Quest. Crops, which protect another crops from trespassing of animals or restrict the speed of wind and are mainly grown as border

Ans. **Border/Guard Crops**

Quest. A crop, grown for direct sale rather than for livestock feed or a crop grown by a farmer primarily for sale to others rather than for his or her own use?

Ans. **Cash Crops**

Quest. Two major commercial crops are

Ans. **(i) Cotton (ii) Sugarcane**

Quest. Crops which are cultivated to catch the forthcoming season when main crop is failed?

Ans. **Catch/Contingent Crops**

Quest. A close-growing crop, grown primarily to improve and protect the soil from erosion through their ground covering foliage and / or rootmats between periods of regular crop production?

Ans. **Cover Crops**

Quest. When both main and intercrop is benefited to each other?

Ans. **Complementary Crops**

Quest. The crops leave the field exhaustive after growing?

Ans. **Exhaustive Crops**

Quest. Any crop or combination of crops is grown for grazing or harvesting for immediate or future feeding to livestock?

Ans. **Ley Crops**

Quest. Such crops are grown to conserve the soil moisture through their ground covering foliage?

Ans. **Mulch Crops**

Quest. The seed of succeeding crops is sown broadcast at 10 to 15 days before harvesting rice crop?

Ans. **Paira/Utera Crops**

Quest. Generally, the third row of crop is removed or growing of crop in pair row and the third row is escaped with an object to conserve the soil moisture in Dryland areas?

Ans. **Paired row Crops**

Quest. Such cops are neither complementary nor competitive?

Ans. **Supplementary Crops**

Quest. Crops, those are grown to protect the main cash crop from a certain pest or several pests?

Ans. **Trap Crops**

Quest. Cereals are botanically

Ans. **Caryopsis**

Quest. The more nutrient exhaustive family is

Ans. **Poaceae (Graminae)**

Quest. The non-conventional oilseed crop is

Ans. **Sunflower**

Quest. The non-edible oilseed crops are

Ans. **Castor and Linseed**

Quest. The Indian originated field crops are

Ans. **Arhar, Mung, Urd, Cotton, Jute, Kodo, Kutki, Oat etc.**

Quest. Kharif crops are generally denoted as

Ans. **Short day plants**

Quest. Rabi crops are generally denoted as

Ans. **Long day plants**

Quest. Day neutral Plants are

Ans. **Cotton, maize, sunflower, safflower, groundnut, buck wheat, tomato.**

Quest. Optimum time of sowing for Kharif crop

Ans. **June-July**

Quest. Optimum time for Rabi crop

Ans. **Last week of October to first week of November**

Quest. Optimum depth of sowing for most of field crops

Ans. **3-5 cm**

Quest. The recommended fertilizer dose (N:P:K) for cereal crops are

Ans. **4:2:1**

Quest. The recommended fertilizer dose (N:P:K) for pulse crops are

Ans. **1:2:1 or 1:2:2**

Quest. The recommended fertilizer dose (N:P:K) for oilseed crops are

Ans. **3:2:1**

Quest. The recommended fertilizer dose (N:P:K) for fodder and fibre crops are

Ans. **2:1:4**

Quest. The C_3 Plants are

Ans. **Rice, Wheat, Barley, Pea, Gram, Mustard and Rye, Cotton, Arhar, Soybean, Sunflower, Lentil, Sugarbeet, Tomato etc.**

Quest. The C_4 Plants are

Ans. **Maize, Sorghum, Bajra, Sugarcane, Millets.**

Quest. The CAM Plants are

Ans. **Pineapple, khajur, cactus, sisal.**

Quest. The optimum temperature for better crop production is between

Ans. **18 – 24^0 C**

Quest. The weight of 1000 seeds of a crop?

Ans. **Test weight**

Quest. The weight of 100 seeds of a crop?

Ans. **Seed Index**

Quest. The net assimilation rate is express in terms of

Ans. **g cm^{-2} day^{-1}**

Quest. Which of the following crop geometry ensures uniform solar radiation availability to crop?

Ans. **Square**

Quest. Wavelength longer than ______ m/µ is not visible to the eye, and are called infrared

Ans. **750**

Quest. The development stage of a plant after which no further increase in dry matter occurs in the economic part is known as

Ans. **Physiological maturity**

(II) Modern Concepts of Tillage

Quest. The mechanical manipulation of the soil is k/s as

Ans. **Tillage**

Quest. The good physical condition of soil after tillage is

Ans. **Tilth**

Quest. Who is the father of tillage?

Ans. **Jethro Tull**

Quest. The tillage operation mainly aims to break, open and turn the soil

Ans. **Primary/tillage**

Quest. The primary tillage implements are

Ans. **Deshi plough, MB plough, Ridge plough, Disk plough etc.**

Quest. The tillage operation, done *to create a good seedbed for proper seeding/planting*

Ans. **Secondary tillage**

Quest. The secondary tillage implements are

Ans. **Cultivator, Harrows, Hoe, Planker, Roller etc.**

Quest. The optimum range of available soil moisture for convenient and effective ploughing

Ans. **50-75%**

Quest. An ideal condition of soil for crop growth?

Ans. **Seed-bed**

Quest. Conventional tillage involves

Ans. **Minimum tillage, Zero tillage and Conservation tillage.**

Quest. The tillage aims to reduce tillage to the minimum necessary for ensuring a good seed-bed, rapid germination, satisfactory stands and favourable growing condition?

Ans. **Minimum tillage**

Quest. The tillage referred as no tillage in which the crop is planted in unprepared soil

Ans. **Zero tillage**

Quest. The word 'Zero tillage' was termed by

Ans. **Jethro Tull**

Quest. A system of tillage in which organic residues are not inverted into the soil and used as a protective cover against erosion and evaporational losses of soil moisture?

Ans. **Conservation tillage or *stubble mulch tillage***

Quest. Conservation tillage tends to encourage

Ans. **Higher microbial population**

Quest. The tillage implement used to break subsoil is

Ans. **Chisel plough**

Quest. Ridge plough is used for

Ans. **Earthing-up and form ridges and furrows.**

Quest. Star weeder is used for

Ans. **Weeding in dry lands and groundnut fields**

Quest. Disc plough is used for

Ans. **Deep ploughing in grassed field**

Quest. Rotary plough is used for

Ans. **Cut and pulverizes the light soil.**

Quest. Harrows are used for

Ans. **Preparation of seedbed, destroy weeds**

Quest. Mechanization index is found highest in

Ans. **Wheat crop**

(III) Cultivation of Field Crops

1. Paddy

Quest. The botanical name and family of paddy

Ans. ***Oryza sativa,*** **Poaceae**

Quest. The chromosome number of paddy

Ans. **2n = 24.**

Quest. The cultivated spp. of paddy

Ans. ***O. sativa and O. glaberima***

Quest. The protein (Oryzein) content in paddy

Ans. **6-7%**

Quest. *Indica* rice is grown in

Ans. **India**

Quest. *Japonica* rice is grown in

Ans. **Japan**

Quest. *Javanica* rice is grown in

Ans. **Indonesia**

Quest. Rice inflorescence is called as

Ans. **Panicle**

Quest. Optimum temperature for good rice crop growth is

Ans. **30-32^{0}C**

Quest. Best pH for cultivation of rice is

Ans. **4-6 pH**

Quest. Sowing of paddy in April-May and harvesting in August-Sept. is called as

Ans. **Aus/Autumn/Pre kharif paddy**

Quest. Sowing of paddy in June-July and harvesting in October is called as

Ans. **Aman/Kharif/Aghani**

Quest. Sowing of paddy in January-Feb and harvesting in April-May is called as

Ans. **Boro/Summer/Spring**

Quest. The best system of rice culture is

Ans. **Transplanting**

Quest. The tillage implement, most suitable for rice cultivation is

Ans. **Power tiller**

Quest. Hulling percentage of rice is

Ans. **70-75%**

Quest. The gene responsible for dwarfness in rice is

Ans. ***Dee-gee-woo-gene***

Quest. First intervarietal cross variety of rice?

Ans. **Jaya (TN1×T141)**

Quest. The rice variety called 'miracle rice' is

Ans. **IR-8**

Quest. Normally rice plant is transplanted at

Ans. **21-25 days after sowing (3-4 leaf stage)**

Quest. Under SRI method, rice plant is transplanted at

Ans. **10-12 DAS (Days After Sowing)**

Quest. SRI denotes

Ans. **System of Rice Intensification**

Quest. In rice 'Dapog seedlings' are ready for transplanting

Ans. **11-14 DAS**

Quest. Dapog method is most commonly prevalent in

Ans. **Philippines**

Quest. The nursery area required for providing seedlings for transplanting 1 ha rice field

Ans. **1000 m^2 (1/10 ha)**

Quest. Most prominent cropping pattern of rice in India?

Ans. **Rice-Wheat**

Quest. Rice prefer nitrogen uptake in

Ans. **Ammonical form (NH^4).**

Quest. The best fertilizer for top dressing in rice?

Ans. **Ammonium sulphate**

Quest. The recommended dose of N, P and K for rice crop

Ans. **100:60:40 kg ha^{-1}**

Quest. For correction of iron chlorosis in rice, following spray is recommended

Ans. **1% solution of ferrous sulphate**

Quest. The nitrogen fixing bacterium found on root surface of rice

Ans. **Azospirillum**

Quest. The most important critical stage of rice for irrigation

Ans. **Tillering to flowering stage**

Quest. In low land rice, fertilizer is applied in

Ans. **Reduced zone only**

Quest. Nitrogen use efficiency in rice is around

Ans. **30-40%**

Quest. Aroma in rice is due to presence of

Ans. ***"Di-acetyl 1 propaline"* chemical**

Quest. Anaerobic environment in rice soil is responsible for gaseous loss of fertilizer nitrogen by

Ans. **Denitrification**

Quest. The Gall midge resistance varieties of rice is

Ans. **Phalguna, Surekha, Suraksha**

Quest. The Blast resistance varieties of rice is

Ans. **Tulsi, IR64**

Quest. The deep water rice are

Ans. **Punkaj, Jagannath**

Quest. Rice varieties suitable for Saline-alkaline soil are

Ans. **CSR-10, CSR-13, CSR-27**

Quest. Super rice variety is

Ans. **Lunishree**

Quest. Gas emitted from rice field is

Ans. **CH_4 (Methan).**

Quest. Weed caused relatively more loss in rice productivity, when it is

Ans. **Direct seeded**

Quest. Most dominated weed species in rice field is

Ans. ***Echinochloa spp.***

Quest. Common herbicide used in rice crop field?

Ans. **Anilophos and Butachlor.**

Quest. Polish percentage of rice is

Ans. 2%

Quest. Khaira disease is caused by

Ans. **Zn deficiency.**

Quest. Akiochi disease is caused by

Ans. **H_2S toxicity.**

Quest. White eye of rice is caused by

Ans. **Fe deficiency.**

Quest. Dead heart and white ear of rice is caused by

Ans. **Yellow stem borer**

Quest. Killer disease of rice are

Ans. **Bacterial Leaf Blight (BLB) and Tungro virus**

Quest. Hydrothermal process of rice which saves vitamin B_{12}?

Ans. **Parboiling**

2. Wheat

Quest. The botanical name and family of wheat

Ans. ***Triticum spp.*, Poaceae**

Quest. Wheat is a

Ans. **Hexaploid plant. (2n = 42).**

Quest. The Mexican dwarf wheat is

Ans. ***T. aestivum* (2n = 42).**

Quest. The bred wheat is

Ans. ***T. vulgare* (2n= 42)**

Quest. Marconi wheat is

Ans. ***T. durum* (2n= 28)**

Quest. Emmer wheat is

Ans. ***T. dicocum* (2n= 28)**

Quest. Indian dwarf/Club wheat is

Ans. ***T. spherococum* (2n= 28)**

Quest. The highest grown wheat species in India

Ans. ***T. aestivum***

Quest. The optimum temperature range for sowing of wheat crop

Ans. **20 to 25 ^{0}C**

Quest. Wheat protein is called as

Ans. **Gluten**

Quest. The protein content in wheat

Ans. **8-11%**

Quest. The flowering portion of wheat

Ans. **Head/Ear/Spike**

Quest. Permanent roots of wheat, appeared after 20-22 days of sowing?

Ans. **Crown roots**

Quest. The shelling percentage of wheat

Ans. **60%.**

Quest. Pearling index in wheat measures

Ans. **Kernel hardness**

Quest. Gene responsible for dwarfness in wheat?

Ans. **Norin**

Quest. Sowing depth of dwarf wheat is directly depend upon

Ans. ***Length of coleoptyle***

Quest. Triple gene dwarf wheat varieties were released during?

Ans. **1970**

Quest. The Row to row spacing of wheat

Ans. **22.5 - 23.0 cm**

Quest. The seed rate of timely sown wheat

Ans. **100 -125 kg/ha**

Quest. The seed rate of Late sown wheat

Ans. **125 -150 kg/ha**

Quest. The most important critical stage of wheat is

Ans. **Crown root initiation (CRI 20-25 DAS)**

Quest. Single gene dwarf varieties are

Ans. **Sonalika, UP-262, WL-711, Girja**

Quest. Double gene dwarf varieties are

Ans. **Kalyansona, UP-215, Arjun, Pratap, Janak**

Quest. Triple gene dwarf varieties are

Ans. **Jawahar, Jyoti, Hira, Moti, Sangam, UP-301, UP-319**

Quest. The variety best suited for sowing in Rainfed areas?

Ans. **C-306, Sujata, Shera, Mukta**

Quest. Marconi wheat varieties are

Ans. **Jayraj, Meghdoot, Malvika, HD-4530 etc.**

Quest. Both blight and Rust resistant varieties are

Ans. **UP 2425, PBW 273, WH 291**

Quest. Most important crop variety during green revolution

Ans. **HD 2329**

Quest. Zinc and sulphates deficiency in wheat field reported in

Ans. **Punjab**

Quest. Most suitable cropping system for wheat crop

Ans. **Mixed cropping**

Quest. Objectionable weed of wheat

Ans. ***Convolvulus arvensis***

Quest. Associated weeds of wheat

Ans. ***Phalaris minor, Avena fatua and Chenopodium album***

Quest. Common herbicide used to control weeds in wheat

Ans. **2, 4-D**

Quest. Initial distinguishing character for identification of *Phalaris minor*

Ans. **It is basal node is pink upto 50 days**

Quest. The moisture content at harvesting stage of wheat

Ans. **25-30%**

3. Maize

Quest. The botanical name of maize is

Ans. ***Zea mays***

Quest. Maize crop is also referred as

Ans. **Queen of cereals**

Quest. The maize protein is known as

Ans. **Zein**

Quest. Protein and oil per cent in Maize grain

Ans. **8–10 per cent and 4–5 per cent**

Quest. Most widely grown maize spp. in India?

Ans. ***Zea mays indurate* (Flint corn)**

Quest. Leading state of rabi maize?

Ans. **Bihar**

Quest. Maize variety widely grown in USA?

Ans. ***Zea mays identata* (Dent corn)**

Quest. The sweetest maize species

Ans. ***Zea mays sacchrata* (Sweet corn)**

Quest. Maize species produce starch similar to tapioca

Ans. ***Zea mays ceretina* (Waxy corn)**

Quest. Seed rate of Hybrid maize

Ans. **20 to 25 kg/ha**

Quest. Seed rate of Composite maize is

Ans. **15 to 20 kg/ha**

Quest. First maize hybrid released in India?

Ans. **1961**

Quest. Single cross technology of maize is given by

Ans. **East and Shull (1910)**

Quest. Double cross technique of maize is given by

Ans. **D.F. Jones (1920)**

Quest. Fodder crop maize varieties are

Ans. **African tall, J1006**

Quest. Pop corn maize varieties are

Ans. **Amber pop, V L Amber. Pop, Pearl pop corn**

Quest. Quality Protein Maize (QPM) varieties released by using

Ans. **Opaqua-2 genes**

Quest. QPM varieties are

Ans. **Sakti, Shaktiman 1 & 2, HQPM**

Quest. Hybrid varieties of maize

Ans. **Ganga-1, 3, 5, 101, Ganga safed-2, Ranjit, Ganga-4**

Quest. Composite varieties of maize

Ans. **Jawahar, Vikram, Kishan, Ambar, Sona, Vijay.**

Quest. The most critical stages for irrigation in maize

Ans. **Silking stage.**

Quest. Maize crop leaves show red and purple colour due to deficiency of

Ans. **Phosphorus (P)**

4. Sorghum/Jowar

Quest. Botanical name of sorghum is

Ans. ***Sorghum bicolor***

Quest. Sorghum crop is also referred as

Ans. **Camel crop**

Quest. The seed rate of sorghum

Ans. **12-15 kg/ha**

Quest. 1st Hybrid variety of sorghum

Ans. **CSH-1 (released in 1965)**

Quest. Alkaloid content present in sorghum leaves

Ans. **HCN (Dhurin alkaloid)**

Quest. Sweet sorghum varieties

Ans. **RSSV 46, 53, 59, 84, 96, NSS 216**

Quest. Varieties suitable for both grain and fodder purpose

Ans. **CSH 13 and CSV 15**

5. Pearlmillet/Bajra

Quest. Botanical name of Pearlmillet

Ans. ***Pennisetum glaucum***

Quest. Pearlmillet is also known as

Ans. **Bulrush millet**

Quest. The seed rate of Pearlmillet

Ans. **5 kg/ha**

Quest. 1st Hybrid variety of Pearlmillet

Ans. **HB-1 in 1965**

Quest. Hybrids varieties of Pearlmillet

Ans. **HB- 1 to 5 and Pusa 23.**

Quest. 80 per cent phosphorus in bajra grains stored in the form of

Ans. **Phytate**

Quest. Productivity of Bajra is highest at

Ans. **UP**

6. Barley

Quest. Botanical name of two rowed barley

Ans. ***Hardium distichoum***

Quest. Six rowed barley is

Ans. ***Hardium vulgare***

Quest. The seed rate of barley is

Ans. **75-80 kg/ha**

Quest. Critical stage for irrigation in barley

Ans. **Active Tiplering Stage (30–35 DAS)**

Quest. 'Pearl Barley' is suited for

Ans. **Kidney disorders**

Quest. Molya disease resistant variety of barley is

Ans. **RD 2052**

Quest. Melting quality is high in variety

Ans. **Rekha**

Quest. Grassy weed in barley field can be effectively controlled by

Ans. **Both Isoproturon and 2,4-D**

Quest. Lugri is a fermented drink developed from

Ans. **Hull less barley grains**

7. Chickpea/Gram

Quest. Botanical name of Desi/Brown Chickpea?

Ans. ***Cicer aeritinum***

Quest. Botanical name of Kabuli/White Chickpea?

Ans. ***Cicer kabulium***

Quest. The most frost affected crop among all field crops?

Ans. **Gram**

Quest. The sour taste in leaf of chickpea is due to presence of?

Ans. **Maleic and Oxalic acid**

Quest. The type of root system in chickpea is

Ans. **Tape root system**

Quest. The requirement of seedbed for better cultivation of chickpea is?

Ans. **Rough seedbed**

Quest. Best soil for cultivation of chickpea?

Ans. **Light alluvial soil (a loose and well aerated soil)**

Quest. The optimum time of sowing of chickpea is?

Ans. **15th to 20th October**

Quest. The seed rate for early sown chickpea is

Ans. **75-80 kg/ha**

Quest. Chickpea variety suitable for rainfed condition?

Ans. **Vishal, Anubhav**

Quest. The early maturing variety of gram is?

Ans. **Chaff chaff, JG-62**

Quest. Chickpea variety resistant to Wilt?

Ans. **JG-74, JG-315, BG-256, Awarodhi**

Quest. Most suitable variety of chickpea to drought resistant?

Ans. **NP-58**

Quest. Late planting of chickpea is done to protect the seedlings from?

Ans. **Wilt disease**

Quest. A process of removal/tipping of apical buds of Chickpea is termed as

Ans. **Nipping**

Quest. The average yield of chickpea in irrigated condition in India

Ans. **12-15 qt/ha.**

8. Pigeonpea/Arhar

Quest. Early maturing pigeonpea is

Ans. ***Cajanus cajan flavus***

Quest. Late maturing pigeonpea is

Ans. ***Cajanus cajan bicolor***

Quest. Pigeonpea belongs to the family of

Ans. **Papilionaceae**

Quest. The type of seed germination in pigeonpea is

Ans. **Hypogeal**

Quest. The sowing time of late maturing pigeonpea is

Ans. **1st week of July**

Quest. The normal seed rate of pigeonpea?

Ans. **10-15 kg/ha**

Quest. Zn deficiency in Pigeonpea is rectified by spraying of

Ans. **5 kg $ZnSO_4$ + 2.5 kg Lime ha^{-1}**

Quest. Extra-short-duration variety of pigeonpea

Ans. **UPAS-120**

Quest. Short duration varieties of pigeonpea?

Ans. **Pusa Ageti, T21, HY2, Pusa 84**

Quest. Sterility mosaic and wilt resistant variety of pigeonpea

Ans. **Amar, Narendra Arhar 1, Azad**

Quest. World's first hybrid variety of Pigeonpea is

Ans. **ICPH-8**

Quest. Harvest index (HI) of Pigeonpea is

Ans. **0.19 (lowest among pulses).**

9. Fieldpea

Quest. Botanical name of Fieldpea

Ans. ***Pisum sativum var. arvense***

Quest. Botanical name of Garden pea

Ans. ***Pisum sativum var. hartense***

Quest. Fieldpea/Grainpea is used for

Ans. **Dal/pulse purpose**

Quest. Gardenpea/Table pea is used for

Ans. **Green pods used for vegetable**

Quest. Seed treatment in pea is done by

Ans. **Captan/Thirum 2.5 gm + *Rhizobium leguminosarum* 10 gm/kg seed**

Quest. The spacing maintained in Fieldpea is

Ans. **30 cm x 5-7 cm**

Quest. The common varieties of Fieldpea

Ans. **Rachana, Arpana, Ambika, T-65,163, Hans, KP-885, Pant C-5**

Quest. Leafless variety of Fieldpea?

Ans. **Arpana**

Quest. The recommended NPK dose for Fieldpeas

Ans. **20:50:30 kg**

10. Mungbean/Greengram

Quest. Latest botanical name of mung is

Ans. ***Phasiolus aureus.***

Quest. The seed rate/ha of mungbean is

Ans. **12-15 kg**

Quest. Early maturing varieties of mungbean is

Ans. **Pusa baisakhi, PS16, K851**

Quest. Yellow Vein Mosaic resistant varieties of mungbean

Ans. **Pant mung 3, Sumrat, Basanti**

Quest. The average yield of mung in India

Ans. **12-15 qt/ha**

11. Urdbean/Blackgram

Quest. Latest botanical name of Urd is

Ans. ***Phasiolus mungo.***

Quest. The seed rate/ha of mungbean is

Ans. **20-25 kg**

Quest. The normal recommended spacing of urdbean

Ans. **40 cm × 10 cm**

Quest. Common varieties of urdbean

Ans. **Pant U-30, JU-2, Type-9, Barkha, Gwalior-2**

Quest. The average yield of Urd in India

Ans. **10-12 qt/ha**

12. Groundnut

Quest. Bunch/Spanish/Erect type groundnut is

Ans. ***Arachis hypogea fastigate***

Quest. Spreading/Verginia runner type groundnut is

Ans. ***Arachis hypogea procumbens***

Quest. Groundnut is a

Ans. **Modified fruit**

Quest. Fruit of groundnut is called

Ans. **Nut**

Quest. The oil and protein content of groundnut

Ans. **45 and 26 %**

Quest. Technology Mission on Pulses and Oilseeds (TEMPO) was started in

Ans. **1986**

Quest. Most suitable soil for groundnut cultivation

Ans. **Sandy loam soil**

Quest. The seed rate of bunch type groundnut varieties

Ans. **100-120 kg/ha**

Quest. The seed rate of Spreading type groundnut varieties

Ans. **100-120 kg/ha**

Quest. The main critical stage of groundnut for irrigation

Ans. **Flowering stage, Pegging stage and Pod formation stage**

Quest. The most suitable irrigation method for groundnut

Ans. **Check basin method**

Quest. The common varieties of bunch type groundnut

Ans. **Jyoti, Kishan, TMV-11, 12, AK-12, 24, Junagarh-11, ICGS-1, 10, 11, 44.**

Quest. The common varieties of Spreading type groundnut

Ans. **Chandra, Type-28, 64, TMV-1, 3, M- 13, 37, Vikram, Verginia, Gangapuri, Godheri-2, 3**

Quest. Earthing-up is done in groundnut crop at

Ans. **35 to 45 DAS**

Quest. Interculture operation in groundnut crop should be avoided at

Ans. **Pegging stage**

Quest. Strain used for biological N_2 fixation in groundnut is

Ans. ***Rhizobium japonicum***

Quest. Vector of virus in groundnut is

Ans. **Aphid**

Quest. Early leaf spot disease of groundnut is caused by

Ans. ***Cercospora arachidicola***

Quest. Late leaf spot disease of groundnut is caused by

Ans. ***Cercospora personata***

13. Sunflower

Quest. The botanical name of sunflower

Ans. ***Helianthus annus***

Quest. Sunflower is also known as

Ans. **Non-conventional oilseed crop**

Quest. Sunflower has high quality edible oil content of

Ans. **45-50%**

Quest. The head of sunflower is called as

Ans. ***Capitulai***

Quest. Best sowing time of rabi sunflower is

Ans. **November 1st to 2nd week**

Quest. The recommended seed rate of sunflower per hectare is

Ans. **5-7.5 kg/ha**

Quest. The recommended spacing between row to row and plant to plant of sunflower?

Ans. **50 cm x 20 cm**

Quest. The most common varieties of sunflower are

Ans. **Modern, MSFH-8, 17, Jwalamukhi, KBSH-1, JS-1, Sunrise selection.**

Quest. The average yield of sunflower in India

Ans. **20-30 qt/ha**

14. Soybean

Quest. The botanical name of soybean is

Ans. ***Glycine max***

Quest. Soybean crop designated as

Ans. ***"Boneless meat"***

Quest. The protein and oil content in soybean seeds

Ans. **40-42%, and 20-22%**

Quest. Soybean is popularly known as

Ans. **Wonder crop**

Quest. Nodule formation in soybean is done by

Ans. ***Rhizobium japonicum***

Quest. The nitrogen fixation per hectare by soybean

Ans. **40 kg**

Quest. The recommended seed rate of soybean

Ans. **75-80 kg/ha**

Quest. The most common varieties of soybean are

Ans. **JS-2, 335, Indira Soya-9, PK-472, 1024, Gaurav, Ankur, Brag, Clark**

Quest. Manturian classified the soybean varieties based on

Ans. **Seed colour**

Quest. Most commonly cultivated soybean in India

Ans. **Yellow coloured soybean**

Quest. The average yield of soybean in India

Ans. **20-25 qt/ha**

15. Rapeseed and Mustard

Quest. The botanical name of Brown/ Indian mustard

Ans. ***Brassica juncea***

Quest. The botanical name of sarson is

Ans. ***Brassica compestris***

Quest. The fruit of mustard is known as

Ans. **Siliqua**

Quest. The recommended seed rate of mustard as main crop is

Ans. **4-6 kg/ha**

Quest. The common varieties of brown sarson

Ans. **Pusa kalyani, Sufla, BSH-1**

Quest. Varieties of mustard are

Ans. **Kranti, Varuna, Krishna, Pusa bold, Vardan, Rohni**

Quest. Hybrid variety Pusa Jai Kisan is also called

Ans. **Bio 902**

Quest. Mustard crop planted at a spacing of 50 x 20 cm will have _______ plants/ha.

Ans. **1,00,000**

Quest. Optimum moisture content for safe storage of mustard is

Ans. **7-8%**

Quest. The critical stages for irrigation in Rapseed and mustard are

Ans. **Rosette stage and Siliqua formation stage**

16. Safflower

Quest. The botanical name of safflower

Ans. ***Carthamus tinctorius***

Quest. Safflower crop is known as

Ans. **Fencing crop/Border crop**

Quest. The oil content in safflower

Ans. **32-36%**

Quest. Fruit of safflower is called

Ans. **Achene**

Quest. The recommended seed rate of safflower

Ans. **15-20 kg/ha**

Quest. The common varieties of brown sarson are

Ans. **JSF-1,2,5, JSI-7, EB-7, JSH-129**

Quest. The average yield of safflower in India

Ans. **18-20 qt/ha**

17. Linseed

Quest. Linseed crop is also known as

Ans. **Flex**

Quest. The botanical name and family of linseed is

Ans. ***Linum usitatisium*, Linaceae**

Quest. The oil percentage in linseed is

Ans. **40-42% oil**

Quest. The linolinic acid present in linseed oil

Ans. **50-60%**

Quest. The recommended seed rate of linseed is

Ans. **25-30 kg/ha**

Quest. The most common varieties of linseed

Ans. **Jawahar-7,17,18, 552, Kiran, Mukta, Sweta, Gourav, Shital**

Quest. Linseed crop require NPK dose of

Ans. **60:40:20 kg ha**

Quest. A process of treatment of stalks for final fibre extraction is termed as

Ans. **Retting**

18. Cotton

Quest. Cotton is popular in America as

Ans. **White gold**

Quest. Indian/old world cottons are

Ans. ***Gossipium arborium, G. herbacium***

Quest. American/new world cotton is

Ans. ***G. hirsutum***

Quest. Egyptian cotton/sea island cotton is

Ans. ***G. barbadence***

Quest. The best soil for cultivation of cotton

Ans. **Black cotton soil**

Quest. The fibre colour of American cotton

Ans. **Creamy White**

Quest. The formulae to calculate ginning percentage

Ans. **Ginning % =**

$$\frac{\textbf{Wt. of lint}}{\textbf{Wt. of seed cotton}} \times 100$$

Quest. The percent of lint in seed cotton is

Ans. **33%**

Quest. Interspecific varieties of cotton

Ans. **Varalaxmi, DCH-32 (hybrid), HB-224, DHB-105**

Quest. Intraspecific varieties of cotton

Ans. **H-4, 6, Savita, Surya (hybrid), JKHY-1**

Quest. Minimum Support Price is fixed by govt. for cotton varieties of

Ans. **H-4, H 777, F 414**

Quest. G-777 is a

Ans. **Indian cotton variety**

Quest. Nitrogenous fertilizer can be top dressed in the cotton up to

Ans. **First flowering**

Quest. The chemical used for delinting of cotton

Ans. $\mathbf{H_2SO_4}$

Quest. Which part of the cotton plant contains lint and fuzz?

Ans. **Hemp**

Quest. 1 bale of cotton is equal to

Ans. **170 kg**

Quest. The average wt. of *Very fine fibre*

Ans. **< 3.0 mg**

Quest. If the fiber length of a cotton hybrid variety is 25 mm, it classified under

Ans. **Long staple cotton**

Quest. Less number of knots in cotton is termed as

Ans. **Superior quality cotton**

Quest. Fibre of cotton contains

Ans. **Cellulose**

Quest. Bt cotton is resistant against the pest

Ans. ***Helicoverpa* (Spotted bollworm)**

19. Jute

Quest. The botanical name of white jute

Ans. ***Corchorus capsularis***

Quest. Jute crop is planted in the month of

Ans. **Feb-March**

Quest. The seed rate of jute per ha

Ans. **8-10 kg/ha**

Quest. Bitterness in jute is due to

Ans. ***Corchorin***

Quest. Low quality of jute fibre attributed to

Ans. **Discolouration of fibre**

Quest. The most common varieties of jute are

Ans. **JRC-321 (Sonali), JRC-212 (Sabuj sona), JRC 7447 (Shyamli), Hybrid C (Padma), KC1 (Joydev) etc.**

Quest. Ideal stage of jute harvesting for fibre purpose

Ans. **Small pod stage/initiation of pod formation (135-140 DAS)**

Quest. Retting of jute fibre is a

Ans. **Biochemical process**

20. Sugarcane

Quest. The botanical name of tropical cane

Ans. ***Saccharum officinarum***

Quest. *Saccharum barberi* and *Saccharum sinensis* are termed as

Ans. **Indian cane**

Quest. Leading state of India in sugarcane production is

Ans. **U.P. (45% of total Prodtn.)**

Quest. Indian Institute of Sugarcane Research (IISR) is situated at

Ans. **Lucknow, Uttar Pradesh**

Quest. Sugarcane Breeding Institute (SBI) is situated at

Ans. **Coimbatore, Tamil Nadu**

Quest. Optimum temperature for sugarcane growth

Ans. **21-27 ^{0}C**

Quest. Inflorescence of sugarcane is called as

Ans. **Arrow**

Quest. The permanent type root of sugarcane is

Ans. **Shoot roots**

Quest. Adsali sugarcane crop planted during the months

Ans. **July–August**

Quest. The requirement of 3 budded sett rate for planting in one hectare land is

Ans. **35,000-40,000 setts**

Quest. Planting material used for sugarcane planting is

Ans. **Upper 1/3 to half part of cane**

Quest. Flat bed method of sugarcane planting is most common in

Ans. **North India**

Quest. Ridge and furrow method is mostly used in

Ans. **South India**

Quest. The chemicals used for sett treatment of sugarcane

Ans. **Agallal and Areton**

Quest. The most critical stage of sugarcane for irrigation is

Ans. **Formative stage (60-130 days after planting)**

Quest. Which bacterium is used for nitrogen fixation in sugarcane field?

Ans. ***Acetobactor diazotrophicus***

Quest. Varieties of sugarcane termed as "Wonder cane"

Ans. **COC-671 and CO-419**

Quest. Earthing up in sugarcane is done at

Ans. **4 month after planting**

Quest. Most commonly used herbicides in sugarcane are

Ans. **Simazine, Atrazine and Alachlor**

Quest. A method of plant analysis for assessing nutrient requirement in sugarcane is

Ans. **Crop Logging**

Quest. Nutrient, responsible for translocation of sugar in sugarcane

Ans. **Potassium (K)**

Quest. The most common symptoms to judge the maturity of sugarcane are

Ans. **Cane become brittle, produces metallic sound and breaks easily at nodes.**

Quest. Sugarcane is considered as mature, when Brix reading is between

Ans. **18-20%**

Quest. Brix reading of juice indicates

Ans. **Total soluble solids**

Quest. Sugar yield from sugarcane is

Ans. **6-10% from juice**

Quest. The sugar content/recovery is more in the cane produced at

Ans. **Southern India**

Quest. By-product of S'cane

Ans. **Molllasses and Baggasses**

Quest. The most dangers disease of sugarcane is

Ans. **Red rot disease**

21. Berseem

Quest. Berseem is also known as

Ans. **Egyptian clover**

Quest. The botanical name of berseem is

Ans. ***Trifolium alexandrinum***

Quest. Sowing of berseem crop is done by

Ans. **Broadcasting**

Quest. The seed rate of berseem is

Ans. **25-30 kg/ha**

Quest. Seed treatment of berseem seed is done by

Ans. ***Rhizobium trifolium* culture**

Quest. First cutting in berseem is done at

Ans. **50-55 days after sowing**

Quest. The popular varieties of berseem

Ans. **Vardan, chindwara, BL-1, 11, 22, 52, C-10 (Maskavi), IGFRI 99-1**

Quest. The objectionable weed of berseem is

Ans. **Kasini (*Chicorium intybus*)**

Quest. Forage yield of berseem crop is

Ans. **800 - 1000 qt/ha**

22. Lucerne/Alfalfa

Quest. The botanical name of lucerne is

Ans. ***Medicago sativa***

Quest. The seed rate of lucerne is

Ans. **20-25 kg/ha**

Quest. Seed treatment of lucerne seeds is done by

Ans. ***Rhizobium meliloti* culture**

Quest. Stem parasitic weed of lucerne is

Ans. ***Cuscuta reflexa* (Doddar)**

Quest. The popular varieties of lucerne

Ans. **Moopa, Rambler, Anand-2, 3, Sirsa 1, 8, 9, Type - 8,9, IGFRI-5,54,244**

Quest. Forage yield of berseem crop is

Ans. **800 - 1100 qt/ha**

Quest. The physiological disorder "Lucerne yellowing" is cause due to the deficiency of

Ans. **Boron (B)**

23. Oat

Quest. The botanical name of oat crop is

Ans. ***Avena sativa***

Quest. The seed rate of oat crop is

Ans. **80 -90 kg/ha**

Quest. Recommended NPK dose for oat

Ans. **80:30:20 kg/ha**

Quest. Best stage for harvesting of oat is

Ans. **Dough stage**

Quest. The popular varieties of oat are

Ans. **Kent, Algerian, UPO 50, Craig, Afterlee, Fulgham, Fleming gold, HFO-114.**

Quest. Total number of cutting taken in oat crop are

Ans. **2-3 cuttings**

Quest. Forage yield of oat crop is

Ans. **400 - 450 qt/ha**

24. Potato

Quest. The botanical name of potato is

Ans. ***Solanum tuberosum***

Quest. The origin place of potato is

Ans. **South America (Peru)**

Quest. The solanin content present in potato is

Ans. **5 mg/100 gm of potato**

Quest. Protein content in potato is

Ans. **1.6%**

Quest. Potato is an

Ans. **Underground stem**

Quest. Potato crop requires an average temperature for tuberization is of

Ans. **17-20^{0}C**

Quest. Potato crop favours the soil for best growth

Ans. **Sandy loam soil**

Quest. Tuber for selected for potato sowing should have

Ans. **At least 3 buds, 2.5-3 cm diameter with 25-30 gm weight.**

Quest. To break the tuber dormancy, the tuber should be treated with

Ans. **1% thiourea + 1 ppm GA_3 for 1 hour**

Quest. The normal seed rate of potato for one hectare land is

Ans. **20-25 qt/ha**

Quest. True potato seed (TPS) enough for planting one ha. crop.

Ans. **40-45 gm**

Quest. The most popular method for potato planting is

Ans. **Ridge and Furrow**

Quest. Earthing-up in potato is done at

Ans. **30 – 45 DAS**

Quest. The short duration varieties of potato is

Ans. **Kufri alankar, Kufri chandramukhi, Kufri bahar, Kufri Jyoti etc.**

Quest. The varieties suitable for late planting of potato is

Ans. **Kufri sinduri, Kufri dewa, Kufri jeevan etc.**

Quest. Most critical stage for irrigation in potato is

Ans. **25% tuber formation stage**

Quest. Dehulming of potato is used to

Ans. **Obtain quality seed tuber by using the chemical $CuSO_4$**

Quest. Seed plot technique (SPT) in potato is used for

Ans. **Producing virus free seed tubers**

Quest. Special size (superior grade tubers) of potato should have

Ans. **≥ 8 cm diameter**

Quest. Potato tubers should be stored in ventilated closed room with maintenance of

Ans. **4-5^{0}C temperature and 90-95% RH**

25. Tobacco

Quest. The botanical name of tobacco is

Ans. ***Nicotiana tabacum***

Quest. *Nicotiana tabacum* is growing for the purpose of

Ans. **Smoking and chewing**

Quest. *Nicotiana rustica* is growing for the purpose of

Ans. **Hookah, chewing and snuff**

Quest. Transplanting age of tobacco is

Ans. **7 to 9 weeks (4-5 leaf stage)**

Quest. Nicotine content (%) of tobacco is

Ans. **0.5 to 5.5 (*N. tabacum*) and 3.5 to 8.0 (*N. rustica*)**

Quest. Cigarette tobacco is prominent growing in states of

Ans. **Andhra Pradesh and Karnataka**

Quest. The seed rate of tobacco is

Ans. **2.5 to 3.0 kg/ha**

Quest. Mutant varieties of tobacco are

Ans. **Jayashri, Bhavya**

Quest. Most critical stages for irrigation of tobacco is

Ans. **Topping**

Quest. As a source of N, potato crop require fertilizer of

Ans. **Potassium nitrate**

Quest. Desuckering of tobacco is done by

Ans. **Melaic Hydracids (2%)**

Quest. Priming method of harvesting is popular in

Ans. **Cigarette, Wrapper and Chewing type**

Quest. Flue curing is done for

Ans. **Cigarette tobacco**

Quest. Fire curing is done for

Ans. **Bidi, Snuff, Chewing, Hookah tobacco**

Quest. Nicotine content accumulates in which part of tobacco?

Ans. **Leaves**

(IV) Cropping and Farming system

Quest. The repetitive cultivation of an ordered succession of crops or crops and fallow on the same land is called as

Ans. **Crop rotation**

Quest. Which of the following rotations is likely to leave soil richer in organic matter?

Ans. **Maize – Oats - Clovers**

Quest. Crop rotation practiced by the majority of the farmers in a given area or locality is called as

Ans. **Cropping Pattern**

Quest. The most prominent and adopted cropping pattern in India

Ans. **Rice - Wheat.**

Quest. The cropping pattern used on a farm and its interactions with farm resources, other farm enterprises and available technology which determine their makeup is called as

Ans. **Cropping system**

Quest. An appropriate combination of farm enterprises *viz.*, cropping system, livestock, poultry, fisheries and the means available to the farmer to raise them for increasing profitability is called as

Ans. **Farming system**

Quest. The raising of animals along with crop production is

Ans. **Mixed farming**

Quest. Growing of two or more crops simultaneously and intermingled without row arrangements, where there is significant amount of intercrop competition is called as

Ans. **Mixed cropping**

Quest. Growing of two or more crops simultaneously in alternate rows or otherwise in the same area, where there is significant amount of inter crop competition is called as

Ans. **Intercropping**

Quest. One crop variety grown alone in pure stands at normal density in a field

Ans. **Sole cropping**

Quest. The repetitive growing of the same sole crop on the same land is termed as

Ans. **Monoculture**

Quest. The growing of more than one crop on the same land in one year is termed as

Ans. **Multiple cropping**

Quest. Growing of two or more crops in quick succession on the same piece of land in a farming year is termed as

Ans. **Sequential/non-overlapping cropping**

Quest. A cropping system where the land is hands over the succeeding crop before the harvest of standing crop.

Ans. **Relay or overlapping cropping**

Quest. Two or more than two crops of different heights cultivated simultaneously on the same field is called as

Ans. **Multistoreyed/Multitired/Multi-level cropping**

Quest. Such crops have different growth habits and zero competition to each other.

Ans. **Parallel cropping**

Quest. The cropping system beneficial to prevent soil erosion due to winds is

Ans. **Strip cropping**

Quest. Cropping intensity of maize-potato-wheat

Ans. **300%**

Quest. The example of parallel cropping is

Ans. **Urd/Moong + Maize**

Quest. Paira and Utera cropping are most probable in

Ans. **Bihar, MP and Chhattisgarh states**

Quest. When the productions of both inter crops is equal to that of its solid planting.

Ans. **Companion cropping**

Quest. Synergetic cropping means

Ans. **When yield of both the crops are higher than their pure crops on unit area *e.g.* Sugarcane + Potato**

Quest. What does 'jhuming' refers to

Ans. **Traditional method of cultivation in hilly area**

Quest. The formulae of cropping intensity is

Ans. **C.I. (%) =**

$$\frac{\textbf{Total cropped area}}{\textbf{Net sown area}} \times 100$$

Quest. The average cropping intensity of India

Ans. **135%**

Quest. The formulae of rotational intensity

Ans. **R.I. (%) =**

$$\frac{\textbf{No. of crops grown in rotation}}{\textbf{Duration of the rotation}} \times 100$$

Quest. "Relative land area under sole crop required to produce the same yield as obtained under a mixed or intercropping system at the same level of management is termed as

Ans. **Land equivalent ratio (LER)**

Quest. Sustainability Yield index value lies between

Ans. **0 to +1**

Quest. The formulae of cropping index is

Ans. **C.I. (%) =**

$$\frac{\textbf{No. of crops grown per year}}{\textbf{Cropped area}} \times 100$$

Quest. Organic farming excludes the application of

Ans. **Fertilizers**

(V) Irrigation Water Management

Quest. The artificial application of water to supply moisture essential to plant growth is termed as

Ans. **Irrigation**

Quest. First entry of water from the upper layer of soil is known as

Ans. **Infiltration**

Quest. Vertical movement of water or downward movement of water from different soil layer is called as

Ans. **Percolation**

Quest. Horizontal flow of water in irrigation channels or through canals is known as

Ans. **Seepage**

Quest. The flow of excess water from the field after saturation of soil.

Ans. **Runoff**

Quest. Downward movement of nutrients and salts from the root zone with the water

Ans. **Leaching**

Quest. Life saving irrigation is also known as

Ans. **Contingency irrigation**

Quest. Volume or quantity of water required for irrigation to bring a crop to maturity

Ans. **Duty of water**

Quest. The total depth of water (cm) required y a crop during its duration in the field

Ans. **Delta**

Quest. The percentage of applied irrigation water stored in the soil and made available for consumptive use by the crop

Ans. **Irrigation Efficiency**

Quest. Soil crusting reduces

Ans. **Infiltration**

Quest. Average annual rainfall of India is about

Ans. **400 Mha.m**

Quest. 75% of rainfall is received by

Ans. **S-W monsoon period (June-Sept)**

Quest. Biggest river basin of India

Ans. **Ganga**

Quest. The quantity of water (gm) necessary for a plant to produce 1 kg of dry matter is known as

Ans. **Transpiration coefficient**

Quest. The process of determining when to irrigate and how much water to apply is termed as

Ans. **Irrigation scheduling**

Quest. Irrigation is applied to the crop at

Ans. **50% soil moisture depletion stage**

Quest. Soil moisture content is determined by

Ans. **Tensiometer (at 0.85 bar)**

Quest. P^F refers to

Ans. **Logarithm of soil moisture tension**

Quest. Which of the following is a method of indirect measurement of soil moisture?

Ans. **Neutron moisture meter**

Quest. Volumetric method of water measurement are

Ans. **Furrows, Sprinklers and Drippers**

Quest. For measuring uniform flow of water, weirs used are

Ans. **Rectangular and Trapezoidal weir**

Quest. Venturi meter is used to measure water, if

Ans. **Water flow from the pipe**

Quest. The most common water flow measuring device which measures water flow in open conduit is

Ans. **Parshall/Venturi flume**

Quest. Irrigation method suitable for lowland rice and jute

Ans. **Flooding**

Quest. Most common method of surface irrigation to irrigate groundnut and pulses

Ans. **Check basin**

Quest. Ring basin method is suitable for

Ans. **Fruit trees**

Quest. The method in which field divided into number of strips by bunds of around 15 cm height is

Ans. **Border strip method**

Quest. The method, suitable for crops *i.e.* Sorghum, cotton, maize, tobacco, potato, sugarcane etc is

Ans. **Furrow method**

Quest. Micro irrigation method, in which water is applied as spray

Ans. **Sprinkler method**

Quest. Irrigation method, suitable for undulating land, sandy soils and Vegetable and fruit crops

Ans. **Sprinkler method**

Quest. Pressure, applied in sprinkler irrigation system

Ans. **>2.5 bar**

Quest. The rate of water delivery in sprinkler system is

Ans. **>1000 litre/hrs.**

Quest. Drip Irrigation is discovered at

Ans. **Israel**

Quest. Drip Irrigation method is also known as

Ans. **Trickle irrigation**

Quest. Micro irrigation method, in which water is applied as drop form through emitters

Ans. **Drip method**

Quest. Drip method is suitable for

Ans. **Wider spaced orchard crops, sugarcane and for saline soils**

Quest. Discharge rate of water per dripper is

Ans. **1-4 litre/hrs**

Quest. The saving of water in sprinkler and drip irrigation methods as compared to surface irrigation methods

Ans. **25-50% and 50-70% water, respectively.**

Quest. Which irrigation method has highest irrigation efficiency?

Ans. **Drip method**

Quest. The water content between Field capacity (-1/3 bar) to PWP (-15 bar) is called

Ans. **Available water**

Quest. The amount of soil moisture or water content held in soil after excess water has drained away is called

Ans. **Field capacity (FC)**

Quest. The moisture content of a soil at which plants permanently wilt and will not recover.

Ans. **Permanent wilting point (PWP)**

Quest. The water, that moves downward freely under the influence of gravity (< 1/3 bar) beyond the root zone

Ans. **Gravitational water**

Quest. The water, retained by the soil in capillary pores (micropores), against gravity (-1/3 to -31 bar) by the force of surface tension

Ans. **Capillary water**

Quest. When water is held tightly as thin film around soil particles by adsorption forces and flows at gravity of > -31 bar, is called

Ans. **Hygroscopic water**

Quest. Capillary movement of water is complemented by

Ans. **Root extension**

Quest. A diffusive process by which liquid water in the form of vapour is lost in the atmosphere

Ans. **Evaporation**

Quest. The process in which soil water lost from leaves of plants in the form of vapour and enters the surrounding atmosphere.

Ans. **Transpiration**

Quest. The quantity of water needed for normal crop growth and yield in a period of time to a place and may be supplied by precipitation or by irrigation or by both.

Ans. **Water requirement of a crop**

Quest. The consumptive use of water is equal to

Ans. **ET + Mw**

Quest. Water requirement of rice is

Ans. **90-250 cm**

Quest. Water requirement of wheat, sorghum, soybean and tobbaco are

Ans. **45-65 cm**

Quest. Water requirement of sugarcane

Ans. **150-250 cm**

Quest. Water requirement of cotton crop

Ans. **70-130 cm**

Quest. Water requirement of maize and groundnut are

Ans. **50-80 cm**

Quest. The period when water requirement is maximum is called as

Ans. **Critical stages of water requirement**

Quest. What will be CPE value when irrigation is scheduled at 0.8 IW/CPE with 6.0 cm depth of irrigation water?

Ans. **7.5 cm or 75 mm**

Quest. The ratio of the crop yield to the total amount of water used for irrigation is called, measured in kg/ha-cm is

Ans. **Water use efficiency**

Quest. Salt content in irrigation water evaluated as best quality

Ans. **0.2 to 0.5 g/lt**

Quest. The permissible and normal limit of EC, RSC, SAR (meq l^{-1}) and Boron content (ppm)

Ans. **2–4, < 2.5, < 10 and <3, respectively.**

Quest. Nitrate levels in drinking water above ______ mg per litre are considered as a human health hazard.

Ans. **10**

Quest. The process of removal of excess water from the field to ensure a favourable salt balance in the soil

Ans. **Agricultural drainage**

Quest. In a waterlogged soil, the concentration of _______ is high

Ans. **Methane**

Quest. The root developed due to water logging in most of the crops

Ans. **Adventitious root**

Quest. Under water logged conditions, which nutrients are found deficient for the crops

Ans. **Both Zn and Cu**

Quest. The depth of water (cm) to be drained in 24 hours period from the entire drainage area.

Ans. **Drainage coefficient**

Quest. Drainage of one ha cm (10^5 lt) in 24 hrs is equal to

Ans. **1.157 litre/sec**

(VI) Watershed Management

Quest. A natural hydrological unit having common runoff outlet point

Ans. **Watershed**

Quest. The ratio of runoff to the volume of precipitation receive in a catchment area is known as

Ans. **Runoff coefficient**

Quest. Micro watershed covers an area of about

Ans. **100 - 1000 ha**

Quest. Major irrigation project covers an area of

Ans. **>10,000 ha**

Quest. Irrigation project covered >10,000 ha of catchments command area (CCA)

Ans. **Major irrigation project**

Quest. Irrigation project covered 2,000 to 10,000 ha of CCA

Ans. **Medium irrigation project**

Quest. Irrigation project covered < 2,000 ha of CCA

Ans. **Minor irrigation project**

Quest. Water harvesting *in situ* is known as

Ans. **Runoff farming**

(VII) Dryland Agriculture

Quest. Cultivation of crops in areas where average annual rainfall is <750 mm per annum

Ans. **Dry Farming**

Quest. Cultivation of crops in areas receiving rainfall from 750 to 1150 mm per annum.

Ans. **Dryland Farming**

Quest. The areas receiving average annual rain fall > 1150 mm are categories as

Ans. **Rainfed Farming**

Quest. Crop growing season of dryland farming is

Ans. **75 – 120 days**

Quest. Change in normal crop planning to meet weather abnormalities is termed as

Ans. **Contingent planning**

Quest. The main important feature of Indian monsoon

Ans. **Long breaks in the rainy season or Prolonged Dry spells**

Quest. The most dangerous situation in dryland condition is

Ans. **Early withdrawal of water**

Quest. The contingent crop plan suggested when long gap in rainfall

Ans. **Providing life saving irrigation only at critical growth stages**

Quest. The alternate crops recommended to sow under late onset of monsoon

Ans. **Castor, Greengram, Cowpea, Sunflower etc.**

Quest. The crop sown under condition of early onset of monsoon

Ans. **Pearlmillet and Sesamum**

Quest. A period in which the available soil moisture is enough to meet the evapotranspiration requirement of dry land crops

Ans. **Length of growing period**

Quest. The length of growing period, suitable for growing only a single dry land crop

Ans. **14 weeks**

Quest. The length of growing period, suitable for inter cropping system

Ans. **14 to 20 weeks**

Quest. The moisture deficit condition, results when the amount of water vapour available in the soil is not sufficient to meet the demand of potential evapo-transpiration

Ans. **Drought**

Quest. The simplest way of adaptation of plant to drought is

Ans. **Evasion**

Quest. The chemical accumulated during drought condition

Ans. **Proline**

Quest. Which is accumulated in the leaves of water stressed plants

Ans. **ABA**

Quest. Most appropriate crops in dryland farming are

Ans. **Pearlmillet, sorghum, gram, toria**

Quest. Which crop rotation under dryland situation will be more remunerative?

Ans. **Sesamum-gram**

Quest. The chemicals used to check transpiration losses of water

Ans. **Antitranspirents**

Quest. 2,4-D, Atrazine and PMA at low concentration act as which type of antitranspirents

Ans. **Stomatal closing**

Quest. Film farming type antitranspirents are

Ans. **Hexadeconal, Mobileaf, Wax and Silicon**

Quest. Reflectant type antitranspirents is

Ans. **Kaoline (5%)**

Quest. Growth retardant type antitranspirents is

Ans. **Cycocel (CCC)**

(VIII) Weed Management

Quest. An unwanted plant, growing where it is not desired?

Ans. **Weed**

Quest. Such weeds, that are grown in cultivated field?

Ans. **Obligate weeds**

Quest. Cropped along with wild land weed are known as?

Ans. **Facultative weeds**

Quest. Problematic weed, whose seed once mixed with crop seed is extremely difficult to separate?

Ans. **Objectionable weed**

Quest. Undesirable, troublesome weed difficult to control

Ans. **Noxious weed**

Quest. Mimicry weeds of rice and wheat is

Ans. ***Phalaris* and wild rice**

Quest. Weed that depends for its growth on its host plant?

Ans. **Parasitic weed**

Quest. Semi root parasitic weed of Sorghum and Sugarcane?

Ans. ***Striga spp.***

Quest. Semi stem parasitic weed of Mango?

Ans. ***Loranthus spp.***

Quest. The total root parasitic weed of Tobacco?

Ans. ***Orabanchi spp.***

Quest. The total stem parasitic weed of Lucerne?

Ans. ***Cuscuta spp***

Quest. The off type crop varieties are

Ans. **Rogue**

Quest. Which of the following stages of a crop are more prone to weed competition?

Ans. **Germination to seedling**

Quest. What is the Critical Period of Crop-Weed Competition for transplanted rice?

Ans. **30-45 DAS**

Quest. What is the Critical Period of Crop-Weed Competition for Upland rice condition?

Ans. **Entire period of crop growth**

Quest. What is the Critical Period of Crop-Weed Competition for sugarcane?

Ans. **30-120 DAS**

Quest. The detrimental effect of one of higher plants on other higher plants is known as

Ans. **Allelopathy**

Quest. The practice of flushing out germinable weed seeds before crop sowing is called

Ans. **Stale seed bed**

Quest. Stale seed bed technique of weed control is a

Ans. **Cultural method**

Quest. 2, 4-D, Simazine, Atrazine and Fluchloralin belongs to the selectivity group

Ans. **Selective herbicides**

Quest. Diquat, Paraquat, Oxadiargyl and Glyphosate etc. belongs to the selectivity group

Ans. **Non-selective herbicides**

Quest. The herbicides applied 1 day before sowing/planting or just are comes under

Ans. **Pre-plant incorporated (PPI) herbicides.**

Quest. The example of PPI herbicides are

Ans. **Fluchloralin, Alachlor, Trifluralin etc.**

Quest. The herbicides applied 1-4 days after sowing are comes under

Ans. **Pre-emergence herbicides.**

Quest. The example of PRE herbicides are

Ans. **Simazine, Atrazin, Alachlor, Butachlor, Nitrofen, Pendimethalin etc.**

Quest. The herbicides applied 30-40 DAS are comes under

Ans. **Post-emergence herbicides.**

Quest. The example of POST herbicides are

Ans. **2, 4-D, Diquat, Paraquat, Isoproturon, Fenoxaprop-ethyl, Sulfosulfuron, Chlorimuron-ethyl etc.**

Quest. 2, 4-D belongs to the chemical group

Ans. **Chloro phenoxy compound**

Quest. Fluchloralin and Pendimethalin belong to the chemical group

Ans. **Dinitroanilines**

Quest. Atazine and Simazine belong to the chemical group

Ans. **Triazines**

Quest. Alachlor, Butachlor and Propanil belong to the chemical group

Ans. **Amide**

Quest. Glyphosate and Anilophos belong to the chemical group

Ans. **Organophosphorus**

Quest. What is the trade name of Alachlor?

Ans. **Lasso**

Quest. What is the trade name of Chlorimuron-ethyl?

Ans. **Classic, Kloben**

Quest. What is the trade name of Chlorimuron 10% + Metasulfuron-methyl 10%?

Ans. **Almix**

Quest. What is the trade name of Ethoxy sulfuron?

Ans. **Sunrise**

Quest. What is the trade name of Glyphosate?

Ans. **Roundup**

Quest. What is the trade name of Nitrofen?

Ans. **Toke E-25**

Quest. What is the trade name of Pendimethalin?

Ans. **Stomp**

Quest. What is the trade name of Fenoxa prop-ethyl?

Ans. **Whip super**

Quest. Paraquate is a

Ans. **Contact herbicide**

Quest. Which herbicide shows Knock down effect?

Ans. **Paraquate, Diquate and Glyphosate**

Quest. Which is a contact selective herbicide?

Ans. **Propanil**

Quest. Herbicides are not used in dust formulation because of

Ans. **Drifting hazards**

Quest. What is the concentration of solution in ppm if 2 kg of 2,4-D is mixed with 1000 lit of water?

Ans. **2000**

Quest. Which of the following weed having herbicide resistance?

Ans. ***Avena fatua***

Quest. Which of the following is a indicator plant for the bioassay of Atrazine?

Ans. **Soybean**

Quest. First biologically controlled weed is

Ans. ***Lantana camara***

Quest. *Parthenium hysteroforus* is biologically controlled by

Ans. ***Zygrogramma bicolarata***

Quest. The most dominant aquatic weed *Eichhornia crassipes* is controlled by

Ans. ***Neochetina bruchi***

Quest. First commercial Bio-herbicide is

Ans. **DEVINE**

Quest. BIPOLARIS is used to control weed

Ans. **Johnson grass**

Quest. Which of the following causes more wastage of herbicide by drift?

Ans. **Ultra-low volume sprayer**

Quest. The Bright Red coloured triangle in herbicide shows

Ans. **Extremely toxic group**

(IX) Soil and Water Conservation

Quest. Detachment and transportation of top soil particles by wind and or by water is known as.

Ans. **Soil erosion**

Quest. What are the types of soil movement in the process of wind erosion?

Ans. **Saltation, Suspension and Surface creep**

Quest. About 50-75% of soil erosion by wind is carried out by

Ans. **Saltation**

Quest. Very fine soil particles (<0.1 mm dia) eroded by mechanism

Ans. **Suspension**

Quest. Removal of soil particles due to rain drops (through bouncing) is called

Ans. **Splash erosion**

Quest. Which mechanism of water erosion is known as "Death of Farmers"?

Ans. **Sheet erosion**

Quest. Chanalization begins from which mechanism of water erosion?

Ans. **Rill erosion.**

Quest. The advanced stage of gully erosion is

Ans. **Ravines**

Quest. Average soil loss million tonnes/ year in India is

Ans. **5,333**

Quest. The land capability classes suitable for crop cultivation are

Ans. **Class I to III**

Quest. According to USDA classification, the land belongs to class VI and VII are suitable for

Ans. **Timber cum fiber farming**

Quest. Agronomical measures are adopted only where land slope is

Ans. **<2%**

Quest. Mechanical measures are adopted only where land slope is

Ans. **>2%**

Quest. Vertical mulches are used only in

Ans. **Black cotton soil**

Quest. The most popular mechanical measure to control soil erosion and conserve is

Ans. **Contour Bunding**

Quest. Contour Bunding is adopted where

Ans. **Land slope (6 %) and in areas where average annual rainfall is < 600 mm.**

Quest. Bench Terracing is practiced on

Ans. **Steep slopping (16-33%) and undulated land**

Quest. The crop grown on degraded land for improvement is called

Ans. **Conservation crop**

Quest. The full form of LEISA is

Ans. **Low External Input Sustainable Agriculture**

Chapter 3

Agrometeorology

Quest. "The study of envelope of air surrounding the planet and of the phenomenon associated with atmosphere."

Ans. ***Meteorology***

Quest. A weather condition over a given region during a longest period.

Ans. **Climate**

Quest. A condition of atmosphere at a given place at a given time.

Ans. **Weather**

Quest. *Monsoon* is a

Ans. **Arabic word**

Quest. Gaseous envelop surrounding the earth known as

Ans. **Atmosphere**

Quest. The ultimate source of energy on the earth is

Ans. **The sun**

Quest. The mean distance between Earth and Sun

Ans. **1.5×10^8 km**

Quest. Temperature on the sun is around

Ans. **6000^0C**

Quest. Who discovered solar energy?

Ans. **Auguste Mouchout**

Quest. The radiation in the sunlight that gives us the feeling of hotness is

Ans. **Infra-red**

Quest. The radiations emitted by the sun and responsible for the cause of skin cancer

Ans. **Ultra-violet**

Quest. In the atmosphere, which of the following gases account for about 99.0 % per cent by volume?

Ans. **Nitrogen, Oxygen, Carbon dioxide**

Quest. Percentage concentration of CO_2 in air and soil is

Ans. **0.030% and 0.25%**

Quest. An average % of solar radiation reaching to the earth

Ans. **50**

Quest. Who is the first scientist attempted to classify the climate?

Ans. **De Condole (1900)**

Quest. Koppen and Thornthwaite classified the climate on the basis of

Ans. **Annual Rainfall and annual Evaporation**

Quest. Troll classified the climate on the basis of

Ans. **Humid month and temperature**

Quest. The instrument able to record almost all meteorological data by desired interval at any time and any place

Ans. **Automatic weather station**

Quest. The value of solar constant is

Ans. **1.94 cal/cm^2/min**

Quest. A certain part of energy received from the sun, is reflected back to space by the earth known as?

Ans. **Albedo**

Quest. The structure of atmosphere is divided on the basis of

Ans. **Vertical temperature difference**

Quest. All weather phenomena (*i.e.* Rain, fog, frost, clouds) occur in the zone of

Ans. **Troposphere**

Quest. Closest and Densest layer of atmosphere

Ans. **Troposphere (8-18 km height)**

Quest. Ozone layer is present in

Ans. **Stratosphere zone (20-48 km)**

Quest. The coldest region of the atmosphere

Ans. **Mesosphere**

Quest. Radio transmission found in

Ans. **Ionosphere zone**

Quest. Gas less zone is

Ans. **Thermosphere (>80 km height)**

Quest. Blue colour of the sky and red colour of sunset is due to

Ans. **Dispersion**

Quest. What is the wavelength of visible solar radiation?

Ans. **0.39-0.7 nm**

Quest. Wavelength longer than ______ m/μ is not visible to the eye, and are called infrared

Ans. **750**

Quest. The weight of the carbon of air at any given place and time.

Ans. **Atmospheric pressure**

Quest. The mean sea level pressure is

Ans. **1013.25 milibars**

Quest. What is wind?

Ans. **Air blowing at a point.**

Quest. North of the equator, surface winds are known as

Ans. **Northeast trade winds**

Quest. Wind direction is determined with the help of instrument

Ans. **Wind vane**

Quest. Wind turbines uses

Ans. **Kinetic energy**

Quest. Heat flow in solid/soil takes place mainly through the process of

Ans. **Conduction**

Quest. Heat flow in liquid/water by process of

Ans. **Convection**

Quest. Heat flow in air by process of

Ans. **Radiation**

Quest. Evaporation is measured by

Ans. **Evaporimeter**

Quest. Wind pressure is measured by

Ans. **Beaufort scale**

Quest. Atmospheric pressure is measured by

Ans. **Barometer**

Quest. Relative humidity (RH) is measured by

Ans. **Hygrometer/Psychrometer**

Quest. Total incoming solar radiation is measured by

Ans. **Pyranometer**

Quest. Evapotranspiration is measured by

Ans. **Lysimeter**

Quest. Combination of Dry bulb and Wet bulb thermometer used for

Ans. **Relative Humidity**

Quest. Rainfall is measured by

Ans. **Raingauge**

Quest. Instrument used for estimating ET under field condition

Ans. **Can Evaporimeter**

Quest. Continuous temperature record by which instrument

Ans. **Thermograph**

Quest. Which Instrument record temperature without contact the object

Ans. **Infrared thermometer**

Quest. Instrument used for measuring concentration of ozone in air

Ans. **Ozonometer**

Quest. Imaginary line that represents the equal temperature

Ans. **Isotherm**

Quest. Lines of equal pressure

Ans. **Isobar**

Quest. Lines of equal rainfall

Ans. **Isohyets**

Quest. Lines of equal cloud cover

Ans. **Isonephs**

Quest. The optimum temperature for better crop production is between

Ans. **18 – 24^0 C**

Quest. Lowest temperature in a day is observed at

Ans. **Just before sunshine**

Quest. What is the dry adiabatic lapse rate in troposphere

Ans. **6.5^{0}C/km**

Quest. The formulae of Relative humidity is

Ans. **RH (%) =**

$$\frac{\text{Water vapour present in the air}}{\text{Water vapour required for saturation}} \times 100$$

Quest. The monsoon covers 75% rainfall in India

Ans. **Southwest monsoon**

Quest. Date of onset of Monsoon in India

Ans. **1st June**

Quest. Date of Monsoon withdrawal in India

Ans. **31st Sept.**

Quest. Average annual rainfall of India

Ans. **400 Mha-m.**

Quest. One particular day, if the rain received 2.5 mm or more

Ans. **Rainy day**

Quest. An average size of rain drop

Ans. **2 mm dia**

Quest. Atmospheric water is known as

Ans. **Green water**

Quest. Soil water is known as

Ans. **Blue water**

Quest. Which clouds are known as rainy clouds?

Ans. **Nimbo-stratus and Cumulonimbus**

Quest. Clouds types which give the heavy and continuous precipitation

Ans. **Cumulonimbus**

Quest. An aggregation of minute drops of water suspended in the air at higher altitude termed as

Ans. **Clouds**

Quest. The unit used to record clouds

Ans. **Okta**

Quest. Cold cloud seeding is done by use of chemical

Ans. **Silver iodide (AgI_2)**

Quest. Warm cloud seeding is done by

Ans. **Sodium chloride (NaCl)**

Quest. Indian Meteorological Organization (IMD) situated in

Ans. **Pune (1932)**

Quest. Phenomenon of warming of eastern pacific

Ans. **EI nino**

Quest. Phenomenon of cooling of eastern pacific

Ans. **LI nino**

Quest. Which surface has least Albedo?

Ans. **Moist black soil**

Quest. An engine of desertification

Ans. **Drought**

Quest. A period of 4 consecutive weeks from May to mid October or 6 consecutive weeks during rest of the year

Ans. **Agriculture drought**

Quest. Widely used index for classification of droughts

Ans. **Palmer drought index**

Quest. The branch of science in which, the collection and interpretation of information about a target without being in physical contact with it?

Ans. **Remote Sensing**

Quest. National Remote Sensing Agency (NRSA) is situated at

Ans. **Hyderabad**

Quest. Medium range weather forecasting is done for

Ans. **3-10 days**

Quest. Forecast will help in planning cropping pattern

Ans. **Long range weather**

Quest. According to Planning Commission, Agro Climatic Zones in India are

Ans. **15**

Quest. According to NBSSLUP, Agro Ecological Regions in India are

Ans. **21**

Quest. The relationship between Celsius and Fahrenheit unit of temperature

Ans. $\frac{C}{5} = \frac{F - 32}{9}$

Quest. A natural warming process involving the interaction of sunlight and carbon dioxide and other gases from the atmosphere

Ans. **Green house effect**

Quest. Three common greenhouse gases include

Ans. **Carbon dioxide, methane, nitrous oxide**

Quest. Chief green house gas responsible for global warming

Ans. **CO_2 (50%)**

Quest. Which green house gas linked with rice crop?

Ans. **Methane (CH_4)**

Quest. The gas, responsible for ozone depletion

Ans. **CF_2Cl_2**

Quest. Which one is a substitute for CFCs

Ans. **Hydrofluorocarbons**

Quest. The chemicals most commonly found in acid precipitation are

Ans. **Sulphuric acid and nitric acid**

Quest. Normal rain water is slightly acidic with a pH of about

Ans. **5.6**

Chapter 4
Soil Science and Biochemistry

(I) Soil Science

Quest. The fine earth covering land surface acts as a reservoir of nutrients and water

Ans. **Soil**

Quest. The word 'Soil' is derived from

Ans. **Latin**

Quest. Study of origin, classification, morphology of soil is known as

Ans. **Pedology**

Quest. The study of soils in relation to crop growth.

Ans. **Edaphology**

Quest. The father of Soil Science

Ans. **Dokuchalev**

Quest. The concentration of soil water in soil

Ans. **50%**

Quest. Organic matter content in Indian soil is

Ans. **5 %**

Quest. Natural soil aggregates/mass are known as

Ans. **Peds**

Quest. The science describes rocks

Ans. **Petrology**

Quest. Granite and Basalt are

Ans. **Igneous rocks**

Quest. Lime stone, Sand stone and Dolomite are

Ans. **Sedimentary rocks**

Quest. Gneiss, Marble, Quartzite and Slate are

Ans. **Metamorphic rocks**

Quest. The rocks gets broken in pieces due to temperature is called

Ans. **Exfoliation**

Quest. Feldspar, Quartz and Mica are

Ans. **Primary minerals**

Quest. Kaolinite, Halloysite and Dickite are

Ans. **1:1 type silicate clay minerals**

Quest. The example of 2 : 1 type silicate clay minerals are

Ans. **Montmorillonite, Vermiculite and Illite**

Quest. Chlorite is

Ans. **2:1:1 or 2:2 type clay mineral**

Quest. The most dominant mineral on earth crust

Ans. **Feldspar (48%)**

Quest. The weathering mineral, having most stable soil structure

Ans. **Kaolinite**

Quest. Which mineral is a source of phosphorus and boron in soils?

Ans. **Apetite**

Quest. The hydroxide act as cementing agent in binding the soil particles together

Ans. **Fe and Al**

Quest. A vertical section of soil through all its horizons

Ans. **Soil Profile**

Quest. Which horizon is called Fertile zone?

Ans. **'A' horizon**

Quest. The horizon absent in arable land

Ans. **'O' horizon**

Quest. The eluviation horizon is

Ans. **'E or A_2' horizon**

Quest. The illuviation horizon is

Ans. **'B' horizon**

Quest. A+B horizons are collectively called as

Ans. **Solum**

Quest. A+B+C horizons together called as

Ans. **Regolith**

Quest. The formulae of bulk density of soil

Ans. **BD (g/cc) =**

$$\frac{\text{Wt. of oven dry soil}}{\text{Volume of soil (Solid + Pores)}}$$

Quest. Bulk density of normal soil is

Ans. **1-1.6 g/cc**

Quest. The formulae of particle density of soil

Ans. **PD (g/cc) =**

$$\frac{\text{Wt. of oven dry soil}}{\text{Volume of soil solid}}$$

Quest. Widely accepted fixed value of particle density is

Ans. **2.65 g/cc**

Quest. The soil having PD 2.50 g/cc and BD 1.25 g/cc will have _____ % porosity.

Ans. **50**

Quest. A field soil sample weighing 60 g, lost12 g on over dying. What is the moisture percent on dry weight basis?

Ans. **25%**

Quest. The weight of one hectare of surface soil (O-15 cm) in kilograms

Ans. **2.24×10^6 kg/ha**

Quest. The arrangement of primary particles of soil

Ans. **Soil structure**

Quest. The relative proportion of sand, silt and clay is termed as

Ans. **Soil texture**

Quest. The best agricultural soil structure is

Ans. **Crumby/Spheroidal**

Quest. Soil structure proving less porosity in soil

Ans. **Platy**

Quest. The best agricultural texture is

Ans. **Loam**

Quest. NBSS and LUP centre is located at

Ans. **Nagpur**

Quest. Commonly followed soil particle classified in India is

Ans. **International Society of Soil Science (ISSS)**

Quest. According to IISS, the particle size of course sand

Ans. **2 - 0.2 mm**

Quest. The particle size of fine sand

Ans. **0.2 -0.02 mm**

Quest. The particle size of silt

Ans. **0.02 – 0.002**

Quest. The particle size of clay

Ans. **< 0.002**

Quest. The maximum pore space are found in

Ans. **Clay soil**

Quest. Soil colour is determined by

Ans. **Munsell Colour chart**

Quest. It is the relative purity or strength of the spectral colour.

Ans. **Chroma**

Quest. The capacity of the soil to change its shape under moist conditions

Ans. **Soil Plasticity**

Quest. The attraction of solid surface for water molecules is called as

Ans. **Adhesion**

Quest. The density of soil water is maximum at

Ans. **4°C**

Quest. The surface tension of water is at 25°C

Ans. **72.7 dyne/cm^2**

Quest. Solution whose strength or concentration is accurately known is termed as

Ans. **Standard solution**

Quest. Water held between 1/3rd and 15 atm

Ans. **Available water**

Quest. The process by which ions are taken into plant roots

Ans. **Absorption**

Quest. The range of usefulness of tensiometer is between

Ans. **0.0-0.8 bar**

Quest. Mechanical analysis of soil is estimated by

Ans. **Stock's law**

Quest. The negative logarithm of H^+ ion concentration

Ans. **Soil pH**

Quest. The pH value varies from

Ans. **0 to 14.00**

Quest. The C : N ratio of the soil are fairly constant between

Ans. **10 : 1 to 12 : 1**

Quest. The C/N ratio in the organic matter of furrow slice (upper 15 cm) of arable soils commonly ranges from

Ans. **8:1 to 15:1**

Quest. The smell of soil after fresh shower is due to

Ans. **Actinomycetes.**

Quest. The most dominant soil order of India

Ans. **Entisol**

Quest. Black soil belongs the soil order

Ans. **Vertisol**

Quest. The soil having more than 30% organic matter is placed in

Ans. **Histosol**

Quest. The most important soil group of India

Ans. **Alluvial Soils**

Quest. Newly formed alluvial soil is called

Ans. **Khadar**

Quest. Which micro-nutrient is most deficient in Indo-Gangatic alluvium soils?

Ans. **Zinc**

Quest. Black soil contains the clay mineral

Ans. **Montmorillonite clay (2:1)**

Quest. The soil deficient in nitrogen content

Ans. **Black soil**

Quest. Black soil shows black colour due to compound

Ans. **Mn**

Quest. The vertical cracks are major problem in

Ans. **Deep black soils**

Quest. Red soil is red coloured due to

Ans. **Ferric oxides**

Quest. Phosphorus fixation is most probable in

Ans. **Laterite Soil**

Quest. Infiltration rate is relatively higher in

Ans. **Sandy soil**

Quest. Peaty soils are generally deficient of

Ans. **Cu**

Quest. Marshy soils are generally deficient of

Ans. **Zn**

Quest. The most deficient micronutrient in the Indian soil is

Ans. **Zn**

Quest. The inherent capacity of the soil to supply nutrients to plants in adequate amount and in suitable proportions

Ans. **Soil Fertility**

Quest. The capacity of the soil to produce plants under a specified programme of management and it is expressed in terms of yields

Ans. **Soil Productivity**

Quest. The process of decomposition of organic matter is termed as

Ans. **Humification**

Quest. Well decomposed FYM contains N, P and K content

Ans. **0.5:0.2:0.5%**

Quest. A mass of rotted organic matter made from waste

Ans. **Compost**

Quest. NPK content of farm compost

Ans. **0.5:0.15:0.5%**

Quest. NPK content of town compost

Ans. **1.4:1:1.4%**

Quest. The organic matter rich compost made by use of earthworms

Ans. **Vermicompost**

Quest. A practice of turning un-decomposed fresh green plant tissue into the soil to improve fertility status and physical structure of the soil.

Ans. **Green Manuring**

Quest. Green manure crops are turned in the field at the stage of

Ans. **Flowering**

Quest. Green manure crops contributes nitrogen ranging from

Ans. **50-175 kg/ha**

Quest. The most widely used green manure crop

Ans. **Sunhemp (*Crotalaria juncea*)**

Quest. The green manure crop having both stem and root nodulation

Ans. ***Sesbania rostrata***

Quest. Green leaf manuring crops

Ans. **Karanj and *Ipomea***

Quest. NPK content of poultry manure

Ans. **3.023:2.63:1.4%**

Quest. The crop oilcake, which has highest nitrification rate

Ans. **Groundnut**

Quest. Groundnut cake contains NPK

Ans. **7:1.5:1.3%**

Quest. Fertilizer which contains only one primary or major nutrient, *e.g.* Urea.

Ans. **Straight fertilizers**

Quest. Those fertilizers having all the three major nutrients *viz.*, N, P and K.

Ans. **Complete fertilizers**

Quest. Fertilizers contain more than 25% of primary nutrients, *e.g.* Urea (46%), DAP (18% N and 46% P_2O_5).

Ans. **High analysis fertilizers**

Quest. The relative percentage of N_2, P_2O_5 and K_2O" in a fertilizer

Ans. **Fertilizer ratio**

Quest. The fertilizer which destroys soil aggregates

Ans. **Sodium nitrate**

Quest. Oldest N fertilizer, best for top dressing in rice

Ans. **Ammonium sulphate (20.6% N and 24% S)**

Quest. Most concentrated nitrogenous fertilizer

Ans. **Anhydrous ammonia (81% N)**

Quest. Explosive fertilizer is

Ans. **Ammoniun nitrate (33% N)**

Quest. Neutral fertilizer, also called Kishan khad

Ans. **Calcium Ammonium Nitrate (26% N)**

Quest. Cheapest N fertilizer, suitable for foliar spray

Ans. **Urea (46% N)**

Quest. Amid form of N fertilizer

Ans. **Urea**

Quest. Which one is considered as organic fertilizer?

Ans. **Urea**

Quest. Water soluble phosphatic fertilizers

Ans. **SSP, DSP, TSP and DAP**

Quest. Citric acid soluble phosphatic fertilizers

Ans. **DCP, Basic slag and Bone meal**

Quest. Oldest commercially available fertilizer

Ans. **SSP (16-18% P_2O_5)**

Quest. The fertilizer which supplies 3 essential plant nutrients

Ans. **SSP**

Quest. Least hygroscopic fertilizer

Ans. **DAP (18% N and 46% P_2O_5)**

Quest. The phosphatic fertilizer suitable for acid soil

Ans. **Bone meal (23-30% P_2O_5)**

Quest. Potassic fertilizer containing highest amount of K_2O

Ans. **Muriate of potash/KCl (60% K_2O)**

Quest. Potassic fertilizer suitable for fertigation

Ans. **Potassium Nitrate (44% K_2O)**

Quest. The medium range of available N in soil

Ans. **280-560 kg/ha**

Quest. Maximum amount of fertilizer is applied in

Ans. **Potato**

Quest. Fertilizer application in lowland paddy is done at

Ans. **Reduced zone**

Quest. Kjeldahl method is used to determine

Ans. **Total N of soil**

Quest. The maximum phosphorus availability in most of the soils is in the pH range

Ans. **6.0 to 6.5**

Quest. The method used to determine available phosphorus from soil is

Ans. **Olsen's method**

Quest. Maximum concentration of urea for foliar spray

Ans. **6%**

Quest. Solution of fertilizer specially applied at initial growth of plants in pulses and vegetable crops

Ans. **Starter solution**

Quest. The application of fertilizer along with irrigation water

Ans. **Fertigation**

Quest. Which nutrients can be applied by fertigation

Ans. **Nitrogen and sulphur**

Quest. The law of diminishing return was proposed by

Ans. **Mitscherlich**

Quest. Lowland applied nitrification inhibitors

Ans. **Oxamide (31% N) and Thiourea (36.8% N)**

Quest. Slowly released N fertilizers

Ans. **S coated urea, Neem coated urea, Urea super granule**

Quest. Symbiotic N_2 fixing bacteria in leguminous crop.

Ans. ***Rhizobium***

Quest. Asymbiotic N_2 fixing bacteria

Ans. **Azotobactor and Azospirillum**

Quest. Rhizobium fixes atmospheric nitrogen/ha to the soil

Ans. **50-100 kg**

Quest. *Azatobactor* can fixes atmospheric nitrogen/ha to the soil.

Ans. **20-30 kg**

Quest. The essential element required by the N fixing bacterium *Rhizobium*

Ans. **Mo**

Quest. Conversion of NH_4 to NO_2 in soil is brought out by

Ans. **Nitrosomonas**

Quest. Conversion of soil nitrate into gaseous nitrogen is

Ans. **Denitrification**

Quest. VAM belongs to the group of

Ans. **Fungi**

Quest. *Rhizobium japonicum* culture is applied for crops

Ans. **Soybean and Groundnut**

Quest. *Rhizobium leguminosarum* is applied for

Ans. **Pea, Lathyrus and Lentil**

Quest. *Rhizobium trifoli* is applied for

Ans. **Berseem**

Quest. In waterlogged rice field, atmospheric nitrogen can be fixed to the soil by

Ans. **BGA**

Quest. The phosphate solubilizer species of micro organisms is

Ans. ***Pseudomonas***

Quest. Mychoryza increase availability of

Ans. **Phosphorus**

Quest. Bio-super is made up of

Ans. **Rock phosphate + Sulpher + Sulphur oxidizing bacteria**

Quest. Zinc solubilizing bacterial biofertilizer is

Ans. **Azozink**

Quest. Total no. of essential nutrients for plants

Ans. **17**

Quest. Total no. of functional nutrients

Ans. **21 (Essential elements + Co, V, Si, Na)**

Quest. Beneficial elements are

Ans. **Ru, Sr, Ni, Cr and As**

Quest. N, P, K, Ca, Mg and S are

Ans. **Macro nutrients**

Quest. N, P and K are

Ans. **Primary nutrients**

Quest. Ca, Mg and S are

Ans. **Secondary nutrients**

Quest. Essentiality of N was established by

Ans. **De Saussure**

Quest. Concept "Essentiality of elements" was proposed in 1939 by

Ans. **Arnon and Stout**

Quest. Arnon and Stout discovered the essentiality of

Ans. **Mo**

Quest. Nutrient uptake both charges form

Ans. **N (NH_4^+ and NO_3^-)**

Quest. Principle uptake form of phosphorus by plants

Ans. **$H_2PO_4^-$**

Quest. Immobile element in soil is

Ans. **Phosphorus**

Quest. Immobile element in plant is

Ans. **Calcium**

Quest. Plant takes the nutrient in the form of

Ans. **Ions**

Quest. The nutrient required for quality maintenance in potato

Ans. **Potassium**

Quest. Nutrient, which maximum uptake by the plants

Ans. **K^+**

Quest. Nutrient essential for oilseed crops

Ans. **S**

Quest. Structural component of Vit B_{12}

Ans. **Cu**

Quest. Deficiency symptoms of N, P, K, Mg and Mo appear in

Ans. **Older leaves**

Quest. New leaves show deficiency symptoms of

Ans. **Fe, Mn, Cu, S**

Quest. Old and new leaves show deficiency symptoms of

Ans. **Zn**

Quest. Deficiency symptoms of Ca and B appear in

Ans. **Terminal buds**

Quest. Cereal crops show 'V' shaped pale yellowing at lower leaf tips due to deficiency of

Ans. **N**

Quest. Deficiency appears as short internodes in plant.

Ans. **N**

Quest. Purple coloration appeared in leaves due to

Ans. **P deficiency**

Quest. Scorching and burning on margins of bottom leaves and irregular fruit development of plant are most common symptoms of

Ans. **K deficiency**

Quest. Failure of terminal bud and root tips is the principle symptom of

Ans. **Ca deficiency**

Quest. Interveinal chlorosis occurs due to

Ans. **Mg and Fe deficiency**

Quest. Downward cupping of leaves in Tobacco and Tea shows

Ans. **S deficiency**

Quest. Whip like structure appeared in terminal bud

Ans. **B deficiency**

Quest. Burning quality of Tobacco decreased due to

Ans. **Chloride**

Quest. Dia back and Little leaf disease in Citrus shows

Ans. **Cu deficiency**

Quest. Mn deficiency shows

Ans. **Interveinal yellowing of younger leaves**

Quest. Translucent spots of irregular shape between veins shows

Ans. **Mo deficiency**

Quest. Upper leaves will show chlorosis on midrib, veins green and dead spots occur in all parts of leaf (veins, tips and margins) show

Ans. **Zn deficiency**

Quest. Zn toxicity is reduced by addition of

Ans. **Superphosphate**

Quest. A situation in which a crop needs more of a given nutrient yet has shown no deficiency symptoms

Ans. **Hidden Hunger**

Quest. Luxury consumed nutrient by plants *i.e.* maize

Ans. **K**

Quest. Excess of N, P and K causes deficiency of

Ans. **Cu**

Quest. Excess of Ca causes deficiency of

Ans. **P**

Quest. Deficiency of N indicated by plant

Ans. **Cauliflower**

Quest. Deficiency of Bo indicated by plant

Ans. **Sugarbeet**

Quest. The soil, which have <4.0 pH is

Ans. **Cat soil**

Quest. The residual effect of urea on soil reaction is

Ans. **Acidic**

Quest. Soil pH > 8.5 indicates soil is

Ans. **Alkaline**

Quest. Saline soil is also called as

Ans. **Solan chalk and White alkali**

Quest. Alkaline soil is also called as

Ans. **Solanetz and Black alkali**

Quest. The saline – alkali soil is also known as

Ans. **Usar**

Quest. Lime ($CaCO_3$) is added to neutralize

Ans. **Acid soils**

Quest. Gypsum ($CaSO_4.2H_2O$) is used for the reclamation of

Ans. **Sodic/Alkaline soils**

Quest. Pyrite (FeS_2) is used to reclaim

Ans. **Saline soils**

Quest. Rock phosphate is applied in

Ans. **Acid soil**

Quest. Gypsum contains

Ans. **29.2% Ca and 18.6% S**

Quest. Method used for the determination of lime requirement of an acid soil is

Ans. **Shoemaker's method**

Quest. The crops prefer acidic soil

Ans. **Rice, tea and potato**

Quest. Highly salt tolerant crops

Ans. **Barley and Sugarbeet**

Quest. Which fertilizer is most beneficial for alkali soils?

Ans. **Ammonium sulphate**

Quest. Which of the plant species can be suggested on saline soil?

Ans. ***Haloxylon salicornium***

(II) Biochemistry

Quest. The word *'Biochemistry'* is a

Ans. **Greek word**

Quest. Father of Agricultural biochemistry

Ans. **Justus von Liebig**

Quest. First used the term 'biochemistry'

Ans. **Neuberg (1903)**

Quest. The most abundant biomolecules on earth

Ans. **Carbohydrates**

Quest. The formulae of monosaccharide

Ans. $C_6H_{12}O_6$

Quest. Monosaccharide contains

Ans. **Glucose, Fructose, Galactose and Mannose**

Quest. Oligosaccharides contains

Ans. **Sucrose, Maltose, Cellobiose, Lactose and Stachyose**

Quest. Monosaccharides possess reducing property due to the presence of

Ans. **Free aldehyde or keto group**

Quest. Glucose is also known as

Ans. **Dextrose**

Quest. Glucose is

Ans. **Corn sugar**

Quest. Biologically active form of glucose

Ans. **D-form**

Quest. Fructose is

Ans. **Fruit sugar**

Quest. Sweetest sugar among all

Ans. **Fructose**

Quest. Monosaccharides that are used as energy source

Ans. **Glucose and fructose**

Quest. Disaccharides contain

Ans. **Sucrose, Maltose, Lactose and Cellobiose**

Quest. Oligosaccharides that is used in preservation of foods

Ans. **Sucrose**

Quest. Trisaccharides contain

Ans. **Raffinose**

Quest. Tetrasacharides contain

Ans. **Stachyose**

Quest. Non reducing type sugar is

Ans. **Sucrose**

Quest. Reducing type sugars are

Ans. **Maltose and Cellobiose**

Quest. Glucose + Galactose consisting of

Ans. **Lactose**

Quest. Sugar presents in milk

Ans. **Lactose**

Quest. Glycogen present in

Ans. **Animal cell**

Quest. Polymer of glucose

Ans. **Cellulose**

Quest. Man can not digest cellulose due to the absence of

Ans. **Cellulose**

Quest. Total no. of essential amino acids are

Ans. **10**

Quest. The term 'Protein' was coined by

Ans. **Moulder (1840)**

Quest. The polymers of amino acid

Ans. **Proteins and enzymes**

Quest. Protein is discovered by

Ans. **Berzeus**

Quest. Protein that contains only amino acid

Ans. **Simple protein**

Quest. Regulatory proteins are

Ans. **Insulin**

Quest. Transport proteins are

Ans. **Haemoglobin and Myoglobin**

Quest. Structural proteins are

Ans. **Collagen and Elastin**

Quest. The most abundant protein present in the world

Ans. **Rubisco**

Quest. Muscle protein is known as

Ans. **Collagen**

Quest. Silk protein is known as

Ans. **Fibrolin**

Quest. Soybean protein is known as

Ans. **Glycinin**

Quest. Wheat protein is known as

Ans. **Gluten**

Quest. Rice protein is known as

Ans. **Oryzein**

Quest. Maize protein is known as

Ans. **Zein**

Quest. Nucleic acids were first discovered by

Ans. **Friedrich Meischer (1868)**

Quest. DNA denotes

Ans. **Deoxyribose nucleic acid**

Quest. Who discovered the base composition of DNA?

Ans. **Chargaff (1953)**

Quest. Deoxyribose sugar + Nitrogenous base is

Ans. **Nucleoside**

Quest. Deoxyribose sugar + Nitrogenous base + phosphate group (PO_2)

Ans. **Nucleotide**

Quest. Single stranded DNA

Ans. **Bacteriophage**

Quest. The form of DNA present in living organisms

Ans. **B-form**

Quest. RNA that transfer amino acids from cytoplasm to ribosome

Ans. **m-RNA**

Quest. Most abundant form of RNA that constitutes 80% parts of cellular RNA

Ans. **t-RNA**

Quest. Which nucleic acid controls all cellular activities?

Ans. **DNA**

Quest. Which nucleic acid is necessary for protein biosynthesis?

Ans. **RNA**

Quest. Double helix model of DNA was proposed by

Ans. **Watson and Crick (1953)**

Quest. The term 'Enzyme' was given by

Ans. **W. Kuhne (1867)**

Quest. Enzyme is discovered by

Ans. **Buckner (1897)**

Quest. 1st enzyme which was discovered by yeast

Ans. **Zymase**

Quest. The enzyme involved in biological nitrogen fixation

Ans. **Nitrogenase**

Quest. Apoenzyme + Prosthetic group

Ans. **Holoenzyme**

Quest. Enzyme without Prosthetic group

Ans. **Apo enzyme**

Quest. Enzymes which exist in multiple forms within single specing of an organism

Ans. **Isoenzyme**

Quest. The non protein component of the enzyme

Ans. **Coenzymes**

Quest. Vitamin was discovered by

Ans. **Funk (1911)**

Quest. Total well defined vitamins are

Ans. **13**

Quest. Vit. B complex (B_1, B_2, B_{12}), B_3, B_6 and Vit. C are

Ans. **Water soluble vitamins**

Quest. Vit. A, D, E and K are

Ans. **Fat soluble vitamins**

Quest. Vitamin A is also known as

Ans. **Retinal**

Quest. Vitamin A deficiency causes

Ans. **Night blindness**

Quest. Vitamin B_1 is also known as

Ans. **Thiamin**

Quest. Deficiency of vitamin B_1 causes

Ans. **Beriberi**

Quest. Vitamin B_2 is also known as

Ans. **Riboflavin**

Quest. Deficiency of vitamin B_2 causes

Ans. **Skin cracking**

Quest. Vitamin B_3 is also known as

Ans. **Pantathenic acid**

Quest. Vitamin B_3 deficiency causes

Ans. **Whiteness of hairs**

Quest. Vitamin B_5 is also known as

Ans. **Niacin**

Quest. Deficiency of vitamin B_5 causes

Ans. **Pellagra**

Quest. Vitamin B_7 is also known as

Ans. **Biotin**

Quest. Deficiency of vitamin B_7 causes

Ans. **Paralysis**

Quest. Vitamin B_{12} is also known as

Ans. **Cynocobalamin**

Quest. Disease caused due to deficiency of vitamin B_{12}

Ans. **Pernicious anaemia**

Quest. Vitamin C is also also known as

Ans. **Ascorbic acid l**

Quest. Vitamin C deficiency causes

Ans. **Scurvy**

Quest. Vitamin D is also known as

Ans. **Calciferol**

Quest. Vitamin D deficiency causes

Ans. **Reckets**

Quest. The vitamin also called Sunshine vitamin

Ans. **Vit. D**

Quest. Vitamin E is also known as

Ans. **Tocopherol**

Quest. Vitamin K is also known as

Ans. **Phyloquinon**

Quest. Disease caused due to deficiency of vitamin E

Ans. **Sterility**

Quest. Disease caused due to deficiency of vitamin K

Ans. **Non-coagulation of blood**

Quest. Cereals are deficient in amino acid

Ans. **Lysine**

Quest. Pulses are deficient in amino acid

Ans. **Methionine**

Quest. Which vitamin contains metal ion?

Ans. **Vitamin B_{12}**

Quest. The reaction of oil/fat with NaOH/KOH as

Ans. **Saponification**

Quest. The number of grams of iodine absorbed by 100 g fat or oil

Ans. **Iodine value/number**

Quest. Value, used to assess the degree of spoilage (rancidity) of a fat or oil

Ans. **Acid Number/Value**

Quest. A partial substitute for petroleum diesel

Ans. **Biodiesel**

Quest. Golden rice is rich in

Ans. **β-carotene**

Chapter 5

Genetics, Plant Breeding, Seed Science and Plant Biotechnology

(I) Genetics

Quest. The study of the way in which genes operate and the way in which they are transmitted (heredity) from parents to offsprings

Ans. **Genetics**

Quest. Who coined the term genetics?

Ans. **W. Bateson (1905)**

Quest. The 'father of modern genetics'

Ans. **Gregor John Mendal**

Quest. Which is known as functional unit of life?

Ans. **Cell**

Quest. Who discovered the cell?

Ans. **R. Hooke (1665)**

Quest. Cell theory was given by

Ans. **M.J. Schleiden and T.Schwann (1939)**

Quest. Which is known as physical basis of life?

Ans. **Protoplasm**

Quest. How many kind of cells are found in living world?

Ans. **2 (Eukaryote and Prokaryote)**

Quest. The plant cell is a type of cell

Ans. **Eukaryote**

Quest. Which cell organelle is found in both prokaryotic and eukaryotic cells?

Ans. **Ribosome**

Quest. The cells without cell wall

Ans. **Protoplasts**

Quest. Controlling centre of cell

Ans. **Nucleus**

Quest. Nucleus was discovered by

Ans. **Robert Brown (1983)**

Quest. Which cell organelle is called as "Power house of the cell"?

Ans. **Mitochondria**

Quest. Who discovered mitochondria?

Ans. **Hollicker**

Quest. Who coined the term mitochondria?

Ans. **Benda (1897)**

Quest. The term 'Endoplasmic reticulum' coined by

Ans. **Porter (1948)**

Quest. Which organelle of cell is known as engine of cell?

Ans. **Ribosome**

Quest. Rough ER is associated with

Ans. **Ribosome**

Quest. The main site of protein synthesis

Ans. **Ribosome**

Quest. 70s type ribosome is found in

Ans. **Mitochondria**

Quest. Name the prokaryotic organism which does not contain mitochondria?

Ans. **Bacteria**

Quest. The term Lysosome was 1st used by

Ans. **Dave (1955)**

Quest. Which organelle of cell is known as suicidal bag of cell?

Ans. **Lysosome**

Quest. The main function of Golgi body is

Ans. **Packing and transport of food materials**

Quest. Which organelle of cell is known as dustbin of cell?

Ans. **Vacuoles**

Quest. Which organelle of cell is non living?

Ans. **Cell wall**

Quest. The material contained in vacuoles

Ans. **Cell sap**

Quest. A self replicating, extra-chromosomal genetic material found in plant cell

Ans. **Plastids**

Quest. Plastids was introduced by

Ans. **Lederberg**

Quest. Which organelles of cell are found only in plants?

Ans. **Plastids, Spherosome**

Quest. Which plastid of cell is responsible for photosynthesis in plants?

Ans. **Chloroplast**

Quest. Which plastid of cell is responsible for colour in plants?

Ans. **Chromoplast**

Quest. Which plastid of cell is responsible for storage of starch and fat in plants?

Ans. **Leucoplast**

Quest. Name the leucoplast which functions as the storage of oil?

Ans. **Lipoplast**

Quest. Thread like bodies that carry gene

Ans. **Chromosome**

Quest. Who firstly discovered chromosome?

Ans. **Strasburger (1875)**

Quest. Who coined the term chromosome?

Ans. **Waldeyer (1888)**

Quest. Who gave the chromosomal theory of Inheritance?

Ans. **Sutton and Boveri**

Quest. What is the fundamental unit of chromosome?

Ans. **Chromatin**

Quest. Which part of the chromosome is known as primary constriction?

Ans. **Centromere**

Quest. The major genetic constituent of chromosome

Ans. **DNA**

Quest. How many daughter cells are formed in one cycle of mitosis?

Ans. **Two**

Quest. 'Mitosis' term was coined by

Ans. **Walter Flemming (1882)**

Quest. In which cells does mitosis occur?

Ans. **Somatic cells**

Quest. Longest phase of mitosis

Ans. **Prophase**

Quest. Shortest phase of mitosis

Ans. **Anaphase**

Quest. The middle stage in which chromosomes are arranged in equatorial plate

Ans. **Metaphase**

Quest. The stage of DNA synthesis in mitosis

Ans. **Interphase**

Quest. How many daughter cells are formed in one cycle of meiosis?

Ans. **Four**

Quest. 'Meiosis' term was 1st given by

Ans. **J.B. Farmer (1905)**

Quest. In which cells does the meiosis occur?

Ans. **Reproductive cells**

Quest. Crossing over and recombination occur during

Ans. **Meiosis division**

Quest. In which stage, crossing over takes place?

Ans. **Pachytene stage**

Quest. *Chiasmata* occurs at

Ans. **Diplotene stage**

Quest. Spindle formation takes place during

Ans. **Metaphase 1**

Quest. The process of separation of chromatids called

Ans. **Disjunction**

Quest. Mendal was born on

Ans. **July 22, 1822**

Quest. Mendal works on the 7 contrasting characters of crop

Ans. **Garden pea**

Quest. Rediscovery of Mendel's work was done by

Ans. **Hugo de Vries, Erich Correns and Erich Tschermak (1900)**

Quest. The accepted theory of Mendal was

Ans. **Law of Segregation**

Quest. Mendelian population is also known as

Ans. **Random mating population**

Quest. A tall pea plant (DD) and a tall pea plant (Dd) have what in common?

Ans. **Phenotype**

Quest. The double helix model of DNA proposed by

Ans. **Watson and Crick (1953)**

Quest. The process of using information (genetic material) from DNA to construct m-RNA

Ans. **Transcription**

Quest. Transfer of genetic material from m-RNA to Protein

Ans. **Translation**

Quest. An expression of one gene depends on the presence or absence of another gene in an individual.

Ans. **Epistasis/gene interaction**

Quest. Phenotypic ratio of Monocross hybrid

Ans. **3:1**

Quest. Phenotypic ratio of Dicross hybrid

Ans. **9:3:3:1**

Quest. Triplet sequence found in mRNA which codes for single amino acid

Ans. **Codon**

Quest. Triplet sequence in t-RNA

Ans. **Anticodon**

Quest. 5 inbred lines will lead to .... no. of single crosses.

Ans. **10**

Quest. The term "Genomics" was coined by

Ans. **Thomas Roderick (1986)**

Quest. Who developed the concept of pangenesis?

Ans. **Darwin**

Quest. Theory of evolution through natural selection was given by

Ans. **C.Darwin and A.R.Wallace (1858)**

Quest. Who developed the theory of acquired character?

Ans. **Lamarck**

Quest. Chromosomal theory of heredity was proposed by

Ans. **W. Sutton (1902)**

Quest. Genes for sex-linked traits are located on

Ans. **"Y" chromosome**

Quest. How many pairs of homologous chromosomes do humans have?

Ans. **23**

(II) Plant Breeding

Quest. The science, which helps in changing the genetic make-up of plants in such a way that they give rise to the maximum economic product for human use.

Ans. **Plant Breeding**

Quest. Basic chromosome/Genomic number is

Ans. **X**

Quest. Genetic chromosome number is

Ans. **n**

Quest. Haploid no. of *Triticum aestivum*

Ans. **n = 21**

Quest. Heteroploid in which one or few chromosomes or missing from 2n

Ans. **Aneuploid**

Quest. Monosomic hypoploid

Ans. **2n -1**

Quest. Double monosomic

Ans. **2n-1-1**

Quest. Nullisomic

Ans. **2n-2**

Quest. Hyperploid have one extra chromosome

Ans. **Trisonic (2n+1)**

Quest. Double trisomic

Ans. **2n+1+1**

Quest. Tetrasomic

Ans. **2n+2**

Quest. Double Tetrasomic

Ans. **2n+2**

Quest. Polyploidy level in embryo and endosperm of seed

Ans. **3n**

Quest. Polyploidy level of testa and pollen mother cell of seed

Ans. **2n**

Quest. Polyploidy level in endosperm of *Triticum aestivum*

Ans. **63**

Quest. Triploids are useful for

Ans. **Seedless fruits**

Quest. Wheat, cotton, tobacco and oat are

Ans. **Allopolyploids**

Quest. The term "primary centres of origin' was proposed by

Ans. **Vevilov**

Quest. Vegetative embryos develops without fertilization

Ans. **Apomixis**

Quest. Development of fruit without fertilization

Ans. **Parthenocarpy**

Quest. Embryo originates from unfertilized egg

Ans. **Parthenogenesis**

Quest. Progeny of a single cross fertilized heterozygous individual

Ans. **Inbred**

Quest. Male sterile line

Ans. **A-line**

Quest. Progeny of a single plant, obtained by asexual reproduction

Ans. **Clone**

Quest. A single gene affecting more than one character/governing multiple traits

Ans. **Pleiotropy**

Quest. Repeated crossing of hybrid progeny back to one of its parents

Ans. **Back Cross**

Quest. F_1 x homozygous recessive parent

Ans. **Test Cross**

Quest. Intra-specific hybridization is a crossed between

Ans. **Two plant of different varieties**

Quest. When pollen grains from an another falls on receptive stigma of the same flowers

Ans. **Self pollination**

Quest. Self pollinated species are also known as

Ans. **Autogamous species**

Quest. When pollen grains from flower of one plant transferred to receptive stigmas of flowers of another plant

Ans. **Cross pollination/Allogamy**

Quest. Rice, Wheat, Buckwheat and Oat are

Ans. **Self pollinated crops**

Quest. Maize, Pearl millet, Black mustard and Sunflower are

Ans. **Cross pollinated crops**

Quest. Often Cross pollination crops are

Ans. **Safflower, Arhar, Cotton and Sorghum**

Quest. Continuous inbreeding (Selfing) leads

Ans. **Homozygosity**

Quest. Single seed descent method is a method of

Ans. **Method of Self pollination**

Quest. Mass selection is always based on

Ans. **Phenotype**

Quest. The oldest selection method of crop improvement

Ans. **Mass selection**

Quest. Concept of pure line was given by

Ans. **Johnson**

Quest. A progeny of single homozygous, self pollinated crops

Ans. **Pure line**

Quest. A method of breeding for wheat

Ans. **Pure line selection method**

Quest. Breeding refers to selection procedure in which the segregating population of self pollinated species is grown without selections

Ans. **Bulk**

Quest. Most commonly used method for selection from segregation generations of crosses in self pollinated crops

Ans. **Pedigree method**

Quest. A method which is not for handling segregating populations

Ans. **Bulk method**

Quest. A method for improving specific traits *i.e.* plant height, disease resistancy

Ans. **Pedigree method**

Quest. Multiline breeding is exploited widely in the crop

Ans. **Wheat**

Quest. A method does not provide opportunity to practice selection for superior plant till F5 generations

Ans. **Single seed descent method**

Quest. Commonly used method for transfer of disease resistancy from one variety to another variety.

Ans. **Back cross method**

Quest. Clonal selection mostly used in the crop

Ans. **Ginger**

Quest. The parent which is used only once in back cross breeding method

Ans. **Donor**

Quest. Bulk method was 1st used by

Ans. **Nilson Ehle (1908)**

Quest. Progeny selection is also known as

Ans. **Ear to row method of selection**

Quest. Methods used for handling the segregating generation

Ans. **Pedigree, Bulk and Single seed descent method**

Quest. A method of breeding is appropriate for improvement of good variety.

Ans. **Back cross method**

Quest. A method in which desirable scattered favourable genes are selected in different plants in each generation

Ans. **Recurrent selection**

Quest. A method of breeding is not appropriate for cross pollinated crops

Ans. **Pure line selection**

Quest. Cross between two genetically different homozygote plants is

Ans. **Hybrid or F_1**

Quest. $A \times B = F_1$

Ans. **Single cross hybrid**

Quest. (A x B) x (C x D)

Ans. **Double cross hybrid**

Quest. Selected variety/line/clone x open pollinated variety

Ans. **Top cross**

Quest. A single cross (A x B) x OP variety

Ans. **Double top cross**

Quest. A variety produced by crossing in all combinations a number of lines that combine well each other

Ans. **Synthetic variety**

Quest. Synthetic variety is maintained by

Ans. **Self pollination**

Quest. A variety produced by mixing the seeds of several phenotypically outstanding lines (varieties)

Ans. **Composite variety**

Quest. Composite variety is developed by

Ans. **Cross pollination**

Quest. Hybrid variety was first exploited in

Ans. **Maize**

Quest. A mechanism of self pollination in which flowers open but only after pollination has take place

Ans. **Chasmogamy**

Quest. Chasmogamy was found in

Ans. **Rice Moong and Oat**

Quest. Pollination and fertilization occurs before opening of flower is termed as

Ans. **Cleistogamy**

Quest. Cleistogamy was found in

Ans. **Wheat and Barley**

Quest. When male and female flowers of a hermaphrodite flower matures at different time

Ans. **Dichogamy**

Quest. When female matures before male

Ans. **Protogyny**

Quest. Pollen from a flower of one plant falls on the stigmas of other flowers of the same plant.

Ans. **Geitonogamy**

Quest. When male and female flowers occur on the same plant

Ans. **Monocius**

Quest. When male and female flowers occur on different plants

Ans. **Diocious**

Quest. Maize is a

Ans. **Monocius plant**

Quest. Papaya is a

Ans. **Diocious plant**

Quest. 1st Intergeneric hybrid was

Ans. ***Raphino brassica* (Radish x Cabbage)**

Quest. Hybrid variety of rice is developed by using

Ans. **GMS and CGMS line**

Quest. Double cross hybrids of maize are developed by

Ans. **CGMS line**

Quest. Exploitation of hybrids in tobacco was carried out by

Ans. **Koelreuter**

Quest. Superiority of F_1 hybrids over both of its parents is termed as

Ans. **Heterosis**

Quest. Term 'Heterosis' was given by

Ans. **Shull**

Quest. The average value for a character of the two parents of the concerned hybrid.

Ans. **Mid parent/Average Heterosis**

Quest. When heterosis estimated over the superior or better parent

Ans. **Heterobeltiosis**

Quest. When superiority of the hybrid to the standard commercial check variety

Ans. **Economic heterosis**

Quest. Exchange of chromatin between non-sister chromatids of homologous chromosomes is known as

Ans. **Crossing over**

Quest. Loss or decrease in vigour and fitness as a result of inbreeding.

Ans. **Inbreeding Depression**

Quest. Highly ID is found in

Ans. **Alfalfa and Carrot**

Quest. Sudden heritable change in any characteristics of an organism

Ans. **Mutation**

Quest. X-rays as mutagen was 1st used by

Ans. **Mullar**

Quest. The unit in which mutation occurs

Ans. **Muton**

Quest. Chemical or physical agent which greatly enhances the frequency of mutation.

Ans. **Mutagen**

Quest. A man made cereal

Ans. ***Triticale***

Quest. The cultivated banana is a

Ans. **Autotriploid**

Quest. Removal of the entire tassel (male inflorescence of maize) from the plant before pollen to initiate cross hybridization

Ans. **Detasseling**

Quest. A condition in which either pollen is absent or non functional in flowering plants.

Ans. **Male Sterility**

Quest. When pollen sterility is caused by cytoplasmic genes

Ans. **Cytoplasmic Male Sterility (CMS)**

Quest. An important male sterility source of sorghum

Ans. **Tift 60**

Quest. Ethrel is used as Gametocide for

Ans. **Wheat, Rice and Sugarbeet**

Quest. Wheat is a

Ans. **Allohexaploid**

Quest. Sugarcane, Cotton and Brassica are

Ans. **Autopolyploids**

Quest. Potato is a

Ans. **Autotetraploid**

Quest. *Brassica nigra* was evolved from

Ans. **B. compestris x B. oleracia**

Quest. *Brassica juncia* was evolved from

Ans. **B. compestris x B. nigra**

Quest. Characters which are governed/ controlled by several genes each having small individual effect.

Ans. **Polygenic Traits**

Quest. Natural genetic engineer

Ans. ***Agrobacterium***

(III) Seed Science

Quest. A fertilized ovule consisting of intact embryo, stored food and seed-coat which is viable and has got the capacity to germinate

Ans. **Seed**

Quest. The seed of a crop variety produced by the breeder which is small in quantity is said to be

Ans. **Nucleus seed**

Quest. Source of breeder seed

Ans. **Nucleus seed**

Quest. Progeny of breeder seed

Ans. **Foundation seed**

Quest. Certified tag colour of breeder seed

Ans. **Golden brown**

Quest. The seeds produced by NSC

Ans. **Foundation seed**

Quest. Certified tag colour of foundation seed

Ans. **White tag**

Quest. Progeny of foundation or registered seed

Ans. **Certified seed**

Quest. Certified seed is generally produced by

Ans. **SSC**

Quest. Certified seed tag having

Ans. **Azure blue colour**

Quest. The emergence and development of seedlings from the seed-embryo under favourable condition

Ans. **Germination**

Quest. Hypogeal germination is found in

Ans. **Cereals, Gram, Arhar, Lentil**

Quest. Epigeal germination is found in

Ans. **Mustard, Sunflower, Castor, Onion.**

Quest. Varietal purity is checked by

Ans. **Grow out test (GOT)**

Quest. The impurity percentage of seed

Ans. **Dockage**

Quest. Formulae of real value of seed

Ans. **Real value of seed =**

$$\frac{\textbf{Purity \%} \times \textbf{Germination \%}}{\textbf{100}}$$

Quest. The capacity of the seed to germinate

Ans. **Seed vigour/viability**

Quest. Seed viability is mostly tested by use of

Ans. **2, 3, 5-triphenyl tetrazolium chloride**

Quest. Formulae of pure live seed

Ans. **Pure Live Seed =**

$$\frac{\textbf{Purity \%} \times \textbf{Viability \%}}{\textbf{100}}$$

Quest. The main aim to maintain isolation distance

Ans. **To avoid contamination or cross pollination**

Quest. Isolation distance for self pollinated crops *i.e.* rice, wheat

Ans. **3 m**

Quest. Isolation distance for certified maize seed

Ans. **200 m**

Quest. Isolation distance for certified pigeonpea

Ans. **50 m**

Quest. Isolation distance for certified sunflower

Ans. **500 m**

Quest. Weight of 1000 seeds is known as

Ans. **Test weight**

Quest. Weight of 100 seeds

Ans. **Seed Index**

Quest. Seed testing refers to

Ans. **Testing of Purity, Moisture and Germination of seeds**

Quest. The standard method of seed moisture estimation

Ans. **Oven dry method**

Quest. The rice variety which has no seed dormancy

Ans. **I.R.-50**

Quest. The dormancy due to hard seed coat or impermeable seed coats

Ans. **Scarification**

Quest. The dormancy due to low temperature and moisture conditions

Ans. **Stratification**

Quest. The PGR used to initiate seed germination

Ans. **Gibberellic acid**

Quest. Seed dormancy of potato tubers is broken by treating tubers with

Ans. **Thiourea 1%**

Quest. The main objective of field inspection is to examine

Ans. **Disease incidence, Isolation distance and Off-types**

Quest. Seed processing is termed as

Ans. **Grading**

Quest. National Seed Corporation (NSC) was registered in

Ans. **1963**

Quest. Seed act was passed on

Ans. **1966**

Quest. The Seed Rule was passed on

Ans. **1968**

Quest. National Seed Project (NSP) was started on

Ans. **1988**

(IV) Plant Biotechnology

Quest. Applied use of molecular biology and recombinant DNA Technology known as

Ans. **Biotechnology**

Quest. The term 'Biotechnology' was coined by

Ans. **Karl Ereky (1919)**

Quest. Multiplication of cell of large number of plants placed in appropriate environment conditions with required nutrients is known as

Ans. **Plant tissue or *in vitro* culture**

Quest. The plant or plant part excised for the *in vitro* cultivation

Ans. **Explant**

Quest. An exact genetic replica of a specific gene or an entire organism.

Ans. **Clone**

Quest. General used nutrient medium in tissue culture

Ans. **B-5 medium and MS medium**

Quest. A culture of isolated mature or immature embryos

Ans. **Meristem culture**

Quest. Young embryo is removed from developing seeds and planted on a suitable nutrient medium *in vitro* is called as

Ans. **Embryo culture**

Quest. Anther or pollen culture technique is used to obtained

Ans. **Haploid plants**

Quest. Culture of an organ *in vitro*

Ans. **Organ culture**

Quest. Capability of an isolated single cell to multiply and differentiate into multicellular organism

Ans. **Totipotency**

Quest. A biochemical process or reaction taking place in a test tube (in lab)

Ans. ***In vitro***

Quest. Alternate forms of a gene

Ans. **Allele**

Quest. A method for transforming DNA especially useful of plant cells

Ans. **Electroporation**

Quest. DNA was 1st synthesized by

Ans. **A. Kornberg (1953)**

Quest. A DNA sequence that codes for a specific polypeptide

Ans. **Cistron**

Quest. A library composed of complementary copies of cellular mRNA

Ans. **cDNA**

Quest. The molecule which encodes genetic information

Ans. **DNA**

Quest. The molecule which helps in decoding genetic information carried by DNA

Ans. **RNA**

Quest. A process of formation of somatic embryos from callus

Ans. **Embryogenesis**

Quest. Crossing of plants through fusion of somatic cell

Ans. **Somatic hybridization**

Quest. A segment of DNA that codes for a specific characters

Ans. **Gene**

Quest. A DNA element which has the ability to move from one chromosomal position to another

Ans. **Jumping gene**

Quest. Father of genetic engineering

Ans. **Paul Berg**

Quest. A bacterium used in genetic engineering

Ans. **E-coli (*Agrobacterium rhizogenes*)**

Quest. Gene responsible for higher amount of lysine in maize

Ans. **Opaque-2**

Quest. PCR denotes

Ans. **Polymerase Chain Reactions**

Quest. RFLD denotes

Ans. **Restriction fragement length polymorphism**

Quest. RAPD denotes

Ans. **Random amplified polymorphic DNA**

Quest. A hybrid produced using nucleous of one parent cell and cytoplasm of both the cell

Ans. **Cybrid**

Quest. Molecular scissors used in genetic engineering

Ans. **Restriction endonuclease**

Quest. Map of genome showing relative positions of genes and or markers on chromosomes

Ans. **Genetic map**

Quest. A single DNA molecule condensed into a compact structure *in vivo* by complexing with accessory histones or histone-like proteins.

Ans. **Chromosome**

Quest. The process of synthesizing multiple copies of a particular DNA sequence

Ans. **Gene cloning**

Quest. The process of producing a protein from its DNA- and mRNA-coding sequences.

Ans. **Gene expression**

Quest. DNA amplification is done in the machine

Ans. **Thermocycler**

Quest. Francis Crick's seminal concept that in nature genetic information generally flows from DNA to RNA to protein.

Ans. **Central Dogma**

Quest. Transplanting a cell, tissue or organ from one nutrient medium to another.

Ans. **Subculture**

Quest. The first biotech plant is

Ans. **Tobacco**

Quest. *Bt* genes are introduced in cotton against the pest

Ans. **Cotton Bollworm**

Quest. The vegetable crop under approval for Bt. technology

Ans. **Brinjal**

Quest. Terminator technology is recently used in

Ans. **Cotton**

Chapter 6
Plant Physiology

Quest. The science concerned with processes, functions, plant responses to external stimulus and growth and development of plant.

Ans. **Plant Physiology**

Quest. Father of plant physiology

Ans. **Stephan Hales**

Quest. The cause of most of the gaseous interchange in soil

Ans. **Diffusion**

Quest. In diffusion, particle/molecules moves from region of

Ans. **Higher to lower concentration**

Quest. The concept 'Osmosis' is given by

Ans. **Abbe Nollel**

Quest. Plasma membrane is a type of

Ans. **Semi-permeable membrane**

Quest. The diffusion of solvent particles into a living cell

Ans. **Endosmosis**

Quest. Strong solution having higher concentration

Ans. **Hypertonic solution**

Quest. Imbibition is coined by

Ans. **Such**

Quest. The component of water potential which determine by the attraction between water and hydrating colloids.

Ans. **Matric potential**

Quest. Shrinkage of protoplasm due to outward flow of water (exosmosis) in a concentrate solution.

Ans. **Plasmolysis**

Quest. The Casparian strip is present in

Ans. **Endosmosis**

Quest. The hydrostatic pressure generated within the cell against cell wall as a result of entry of water into it, due to osmosis.

Ans. **Turgor pressure**

Quest. During Osmosis, movement of water takes place from

Ans. **Lower to higher concentration of solution**

Quest. Water is absorption by plants mainly through

Ans. **Root hairs**

Quest. First step in absorption of water is

Ans. **Imbibition**

Quest. Absorbed of water against a concentration gradient by using energy released from respiration is called as

Ans. **Active absorption**

Quest. Transpiration associated ion uptake is

Ans. **Passive uptake process**

Quest. Nutrients absorbed by plants from soil solution are carried upward through the

Ans. **Xylem**

Quest. Downward movement of food synthesized in leaves takes place through

Ans. **Phloem**

Quest. The movement of nutrient ions and salts along with moving water

Ans. **Mass flow**

Quest. Mass flow or pressure flow theory was given by

Ans. **Godlewski (1884)**

Quest. Pulsation theory was given by

Ans. **J.C. Bose (1923)**

Quest. The hydrostatic pressure developed due to the accumulation of water absorb by the root is called root pressure.

Ans. **Root pressure**

Quest. Root pressure is measured by

Ans. **Manometer**

Quest. Upward translocation of fluid in xylem takes place due to

Ans. **Pull of transpiration stream**

Quest. The most accepted theory of water absorption

Ans. **Transpiration pull theory**

Quest. Plant cell walls are

Ans. **Permeable in nature**

Quest. Minerals are translocated in plants as

Ans. **Both organic and inorganic compounds**

Quest. The plant meet their carbon requirement by absorbing

Ans. **CO_2 for atmosphere**

Quest. The process by which plants convert light energy of photon (captured from sunrays) into chemical energy

Ans. **Photosynthesis**

Quest. The oxidation reduction process is

Ans. **Photosynthesis**

Quest. Photosynthesis active radiation (PAR) occurs at

Ans. **400-700 nm**

Quest. Plant component responsible for photosynthesis is a pigment called

Ans. **Chlorophyll**

Quest. Chlorophyll contains

Ans. **Mg**

Quest. Pigment which are responsible for photosynthesis in higher plants

Ans. **Chlorophyll a and b**

Quest. The colour of chlorophyll a

Ans. **Blue green**

Quest. The colour of chlorophyll b

Ans. **Yellow green**

Quest. Oxygen required for photosynthesis comes from

Ans. **Water**

Quest. The product of photosynthesis which is used for growth and development of plants

Ans. **Glucose**

Quest. Photosynthesis can be measured by measuring

Ans. **O_2 given off and CO_2 uptake**

Quest. Chemicals which retard transpiration rate called

Ans. **Anti-transparent**

Quest. Which organism/groups do not have photosynthesizing capability?

Ans. **Fungi**

Quest. The first biological process that begins in a seed soon after in imbibes water

Ans. **Respiration**

Quest. C_4/Light reaction/Hill reaction take place in

Ans. **Grana of Chlorophyll**

Quest. C_3/Dark reaction/Calvin cycle take place in

Ans. **Stroma of Chlorophyll**

Quest. Grana and Stroma are found in

Ans. **Chloroplast**

Quest. C_3 cycle is also known as

Ans. **Blackman reaction**

Quest. Final product of C_3 pathway is

Ans. **3 PGA**

Quest. C_3 plants are

Ans. **Rice, Wheat, Pea, Soybean etc**

Quest. The most important enzyme involved in photosynthetic CO_2 fixation in C_3 plants

Ans. **Rubisco**

Quest. C_4 cycle is also known as

Ans. **Hatch and slack pathway**

Quest. First product of C_4 pathway is

Ans. **Oxalo acetic acid (OAA)**

Quest. C_4 plants are

Ans. **Maize, Sorghum, Sugarcane, millets etc.**

Quest. The most important enzyme involved in photosynthetic CO_2 fixation in C_4 plants

Ans. **PEP carboxylase**

Quest. C_4 plants have

Ans. **Kranz type leaf**

Quest. CAM denotes

Ans. **Crassulacean Acid Metabolism**

Quest. The example of CAM plants are

Ans. **Pineapple and Opuntia**

Quest. The water use efficiency of C_4 is …….. than C_3 plants

Ans. **High**

Quest. The photosynthetic rate of C_4 is …….. than C_3 plants

Ans. **Low**

Quest. C_4 plants normally give more biological yield than C_3 plants because of

Ans. **Less respiration**

Quest. One $NADH_2$ produces how many ATP molecules?

Ans. **3**

Quest. Oxygen is required by the plants for

Ans. **Respiration**

Quest. Glycolysis occurs in

Ans. **Cytoplasm**

Quest. Final product of Glycolysis

Ans. **Pyruvate**

Quest. Net gain ATP during glycolysis

Ans. **2**

Quest. Kreb cycle and ETC occurs in

Ans. **Mitochondria**

Quest. Net gain ATP synthesis from one molecule of glucose in respiration

Ans. **36 ATP**

Quest. One molecules of ATP yields

Ans. **7.6 Kcal energy**

Quest. An energy spending process

Ans. **Photorespiration**

Quest. Photorespiration occurs in

Ans. **Night**

Quest. Photorespiration occurs only in

Ans. **Chlorophyllous cells**

Quest. The loss of water in the form of vapour from the living aerial parts of the plant is known as

Ans. **Transpiration**

Quest. The principle organ of transpiration is

Ans. **Stomata of leaf (90% transpiration)**

Quest. The water is lost during transpiration in the form of

Ans. **Vapour**

Quest. Stomata is found mainly on

Ans. **Lower surface of leaves**

Quest. Opening and closing of stomata are due to its

Ans. **Turgidity and faccidity**

Quest. The types of stomata mostly present on lower surface of leaves

Ans. **Potato type**

Quest. Stomata that is present on only under surface of leaf.

Ans. **Apple and Mulberry type**

Quest. The loss of water (contains salts and minerals) through hydathodes in liquid form during night and regulated by root pressure

Ans. **Guttation**

Quest. The irreversible change in any plant part (s) with respect to size, form, weight, volume etc.

Ans. **Growth**

Quest. The phasic change of individual cells into tissues, organs and organisms

Ans. **Development**

Quest. The growth rate of plants is measured by

Ans. **Auxanometer and Crescograph**

Quest. The element which takes part in the growth and development of plants

Ans. **Plant nutrients**

Quest. Organic compounds which inhibit or modify any physiological process

Ans. **Plant Growth Regulators (PGRs)**

Quest. Growth promoters are

Ans. **Auxins, Gibberellins and Cytokinin**

Quest. Example of growth inhibitors

Ans. **Abscisic acid and Ethylene**

Quest. The PGR causes apical bud dominance

Ans. **Auxins**

Quest. The senescence is delayed by

Ans. **Cytokinin**

Quest. Seed dormancy of seed is broken by

Ans. **Cytokinin**

Quest. Seed dormancy is induced by

Ans. **Abscisic acid**

Quest. PGR used for fruit ripening

Ans. **Ethylene**

Quest. Abscisic acid is synthesized from

Ans. **Actively growing points**

Quest. Cytokinin is mainly synthesized in

Ans. **Root tips**

Quest. PGR related to drought tolerance and stress hardness in plants

Ans. **Abscisic acid**

Quest. The physiological response of plants in relation to length of light

Ans. **Photoperiodism**

Quest. Short day plants require day length

Ans. **<10 hrs**

Quest. Kharif crops requires

Ans. **Shorter day length**

Quest. Generally rabi crops are

Ans. **Long day plants**

Quest. Long day plants require day length

Ans. **>14 hrs**

Quest. The sites of Vernalisation

Ans. **Apical buds/Growing point**

Quest. For germination, seed depends on external source for supply of

Ans. **Water**

Quest. The temperature at which highest percentage of seed germination occurs in short period of time

Ans. **Optimum temperature**

Quest. Ca is essential for

Ans. **Cell wall formation**

Quest. The region of plants in which food to be translocated originates

Ans. **Source**

Quest. The region of plants in which translocated food is utilized or immobilized

Ans. **Sink**

Quest. Concentration of a nutrient in plant tissue where growth of the plant is slowed down

Ans. **Critical concentration**

Chapter 7

Major Pest and Diseases of Important Crops

(1) Pest of Rice

Quest. The scientific name of Yellow Stem Borer

Ans. ***Scirphophaga incertulas***

Quest. The scientific name of Rice Gall Midge

Ans. ***Orseolia oryzae***

Quest. The scientific name of Green Leaf hopper (GLH)

Ans. ***Nephotetrix nigropictus***

Quest. The scientific name of White-backed Plant-hopper (WBPH)

Ans. ***Sogatella furcifera***

Quest. The scientific name of Brown Plant hopper (BPH)

Ans. ***Nilaparvata lugens***

Quest. Which pest causes severe damage to rice panicle at night?

Ans. **Rice Army worm (*Mythimna saparata*)**

Quest. Chaffy grains with black spot is the infestation of

Ans. **Gundhi Bug (*Leptocorisa acuta*)**

Quest. Clipping off the top of rice seedlings containing immature stages of insects reduces the carry over of infestation of

Ans. **Rice hispa**

(2) Wheat

Quest. The scientific name of Pink Stem Borer

Ans. ***Sesamia inferens***

Quest. Pink stem borer attacks to plants in

Ans. **Night**

Quest. The scientific name of Wheat Termites

Ans. ***Odentotermis obesus***

Quest. The scientific name of Cut Worm

Ans. ***Agrotis ipsilon***

Quest. Which pest attacks all the parts of the plant?

Ans. **Termite**

(3) Sorghum

Quest. The scientific name of Sorghum Shoofly

Ans. ***Atherigona varia soccata***

(4) Chickpea

Quest. The scientific name of Cut worm

Ans. ***Agrotis ipsilon***

Quest. The scientific name of Pod borer

Ans. ***Helicoverpa armigera***

(5) Pigeonpea

Quest. The scientific name of Pod borer

Ans. ***Etiella zincknella***

Quest. The scientific name of Plume moth

Ans. ***Exelastis atomosa***

Quest. The scientific name of Pod fly

Ans. ***Melanagromyza obtuse***

Quest. The scientific name of Pod bug

Ans. ***Clavigralla gibbosa***

(6) Soybean

Quest. The scientific name of Soybean Girdle beetle

Ans. ***Oberea brevis***

Quest. The scientific name of stemfly

Ans. ***Melanagromyza sojae***

Quest. Which pest is known as stem borer of soybean

Ans. **Girdle beetle**

(7) Groundnut

Quest. The scientific name of Groundnut aphid

Ans. ***Aphis craccivora***

Quest. The scientific name of White Grub of groundnut

Ans. ***Holotrachia conseguina***

Quest. The scientific name of Groundnut Leaf minor

Ans. ***Stomoperyx nertaria***

(8) Mustard

Quest. The scientific name of Mustard aphid

Ans. ***Lipaphis erysimi***

Quest. The scientific name of Mustard sawfly

Ans. ***Athaliya proxima***

Quest. The scientific name of Mustard Painted Bug

Ans. ***Bargrada cruciferarum***

(9) Sunflower and Linseed

Quest. The scientific name of Capitulai/Head borer

Ans. ***Helicoverpa armigera***

(10) Cotton

Quest. Rosetting of flowers and double seed formation is the symptoms of

Ans. **Pink Bollworm (*Pectinophora gossypiella*)**

Quest. The pest causing flaring of squares in cotton

Ans. **Spotted Bollworm (*Earias vitella*)**

Quest. The scientific name of American Bollworm

Ans. ***Helicoverpa armigera***

Quest. The main symptom of American bollworm is

Ans. **Larger circular bore holes with faecal pellets.**

Quest. The vector of leaf curl virus

Ans. **White Fly (*Bemisia tabaci*)**

Quest. The scientific name of Red Cotton Bug

Ans. ***Dysdercus koenigii***

Quest. Which cotton pest causes hopper burn?

Ans. **Cotton Jassid (*Amrasca bigutulla*)**

Quest. Curling of leaf upwards and yellowing of terminal cotton shoots is a characteristics symptom of presence of

Ans. **Cotton Aphid**

Quest. Highest consumption of pesticides found in

Ans. **Cotton (54%)**

(11) Sugarcane

Quest. The scientific name of Sugarcane Pyrilla/Leaf hopper

Ans. ***Pyrilla purpusilla***

Quest. Which pest causes Bunchi top appearance in sugarcane

Ans. **Top Borer (*Tryporza novella*)**

Quest. The scientific name of Sugarcane shoot borer

Ans. ***Chilo sacchariphagus***

Quest. The scientific name of Sugarcane root borer

Ans. ***Emmalocera depressella***

Quest. The scientific name of Whitefly of sugarcane

Ans. ***Aleurolobus barodensis***

(12) Potato and Tobacco

Quest. The scientific name of Potato aphid

Ans. ***Aphis gossypii***

Quest. The scientific name of Potato tuber moth

Ans. ***Phthorimaea operculella***

Quest. The scientific name of Tobacco cutworm

Ans. ***Spodoptera litura***

(13) Pest of Fruit Crops

Quest. The scientific name of Mango hopper

Ans. ***Amritodus atkinsoni***

Quest. The scientific name of Mango mealy bug

Ans. ***Drosicha mangiferae***

Quest. Sticky bands around tree trunks provide protection against

Ans. **Mango mealy bug**

Quest. The scientific name of Banana Stem weevil

Ans. ***Odoiporus longicollis***

Quest. The scientific name of Guava fruitfly

Ans. ***Bactrocera diversus***

Quest. The scientific name of Fruit sucking moth of citrus

Ans. ***Otheris materna***

Quest. The scientific name of Lemon butterfly

Ans. ***Papillio demoleus***

Quest. The scientific name of Citrus Psylla

Ans. ***Diaphorina citri***

Quest. The scientific name of Papaya fruitfly

Ans. ***Bactrocera/Dacus dorsalis***

Quest. The scientific name of Fruit Borer of pomegranate

Ans. ***Conogethes punctiferalis***

Quest. The scientific name of Woolf aphis of apple

Ans. ***Eriosoma lanigerum***

(14) Pest of Vegetable Crops

Quest. The scientific name of Tomato fruit borer

Ans. ***Helicoverpa armigera***

Quest. The scientific name of Brinjal fruit and shoot borer

Ans. ***Leucinodes orbonalis***

Quest. The scientific name of fruit and shoot borer of Okra

Ans. ***Earias vitella***

Quest. The scientific name of Chilly thrips

Ans. ***Thrips tabaci***

Quest. The scientific name of Red Pumpkin beetles of cucurbits

Ans. ***Raphidopalpa foveicollis***

Quest. The scientific name of Cucurbits fruitfly

Ans. ***Dacus cucurbitae***

Quest. The scientific name of Diamond back moth (DBM) of cabbage

Ans. ***Plutella xylostella***

Quest. The scientific name of Cabbage head borer

Ans. ***Hellula undalis***

(15) Stored Grain Pests

Quest. The scientific name of Khapra beetle or Wheat beetle

Ans. ***Trogoderma granarium***

Quest. The scientific name of Red flour beetle

Ans. ***Tribolium castaneum***

Quest. The scientific name of Pulse beetle

Ans. ***Callosobruchus chinensis***

Quest. The scientific name of Rice moth

Ans. ***Corcyra cephalonica***

DISEASES OF IMPORTANT CROPS

(1) Rice

Quest. The causal organism of brown spot of rice

Ans. ***Helminthosporium oryzae***

Quest. The brown spot of rice is

Ans. ***Externally seed borne disease***

Quest. Poor man's disease of rice

Ans. **Bacterial leaf blight (*Xanthomonas oryzae*)**

Quest. The most destructive phase of the bacterial blight of rice is known as

Ans. **Kresek**

Quest. The air borne disease of rice

Ans. **Rice Blast (*Pyricularia oryzae*)**

Quest. Rice blast is effectively controlled by spraying of

Ans. **Edifenphos**

Quest. The soil borne disease of rice

Ans. **Sheath blight of rice (*Rhizoctonia oryzae*)**

Quest. Khaira disease of rice is caused by

Ans. ***Zinc deficiency***

Quest. Khaira disease of rice is controlled by spraying

Ans. **Zinc sulphate (5 kg)+ lime (2.5 kg/ha) in 10 days nursery**

Quest. The main symptom of Tungro disease of rice

Ans. **Yellowing of leaves**

Quest. The vector of Tungro disease

Ans. **GLH**

Quest. The disease responsible for the great Bengal famine in 1942-43

Ans. **Brown spot of rice**

Quest. Montek disease of rice is caused by

Ans. **Rice root nematode**

(2) Wheat and Barley

Quest. The soil, air and seed borne disease of wheat

Ans. **Kernal bunt (*Neovossia indica*)**

Quest. The foul smell of kernel bunt infected field is due to

Ans. **Trimethyl-amine**

Quest. Loose smut of wheat is a

Ans. **Internally seed borne disease**

Quest. Loose smut of wheat is caused by

Ans. ***Ustilego nuda triticii***

Quest. Loose smut of wheat can be controlled by seed treatment with

Ans. **Vitavax**

Quest. Solar heat treatment is used to control

Ans. **Loose smut of wheat**

Quest. Which rust was earliest appeared in India on wheat

Ans. **Brown/Orange/Leaf rust**

Quest. Alternate host of black stem rust of wheat in India is

Ans. ***Berberries* sp.**

Quest. The causal organism of Black/ stem rust

Ans. ***Puccinia gramini triticii***

Quest. The causal organism of Yellow/ strip rust

Ans. ***Puccinia striformis***

Quest. The causal organism of Brown/ orange/leaf rust

Ans. ***Puccinia recondita***

Quest. Which disease of wheat is discovered in Haryana?

Ans. **Kernal bunt**

Quest. Bacterial rot of wheat ears is also known as

Ans. **Spike blight/Tundu/Yellow slime disease**

Quest. The causal organism of covered smut of barley

Ans. ***Ustilago hardei***

Quest. Sooty or charcoal like powdery mass usually appearing on floral organs particularly the ovary is

Ans. **Smut**

Quest. Which stage of the wheat rust fungus is considered as the perfect stage ?

Ans. **Telial stage**

(3) Maize, Sorghum and Bajra

Quest. White bud of maize is caused by

Ans. ***Zn deficiency***

Quest. The causal organism of Grain Smut of sorghum

Ans. ***Sphacelotheca sorghi***

Quest. Grain Smut of sorghum is also known as

Ans. ***Covered/Kernel/Shoot smut***

Quest. Grain and Head smuts are

Ans. **Seed borne disease**

Quest. The causal organism of Head Smut of sorghum

Ans. ***Sphacelotheca relliana***

Quest. The most serious smut among the smuts affecting sorghum is

Ans. **Grain smut**

Quest. Downey Mildew/Green Ear disease of Bajra is caused by

Ans. ***Sclerospora graminicola***

Quest. Ergot disease of Bajra is caused by

Ans. ***Claviceps fusiformis***

Quest. Ear showing honey dew symptoms is characteristic feature of

Ans. **Grain smut of bajra**

Quest. Smut disease infect the plant at

Ans. **Tillering stage**

(4) Chickpea and Pigeonpea

Quest. Wilt disease is a

Ans. ***Soil borne disease***

Quest. Wilt disease of chickpea is caused by

Ans. ***Fusarium oxisporium***

Quest. Wilt of pigeonpea is caused by

Ans. ***Fusarium oxisporum f. sp. udum***

Quest. The vector of sterility mosaic of pigeonpea

Ans. **Mite *(Aceria cajani)***

(5) Groundnut

Quest. Leaf spot of groundnut is also known as

Ans. **Tikka disease**

Quest. Early leaf spot of groundnut is caused by

Ans. ***Cercospora arachidicola***

Quest. Late leaf spot of groundnut is caused by

Ans. ***Cercospora parsonata***

Quest. The causal organism of collar rot disease

Ans. ***Aspergillus niger***

Quest. Rust of groundnut is caused by

Ans. ***Puccinia arachidis***

Quest. Vector for bud necrosis of groundnut is

Ans. **Thrips**

(6) Rapseed and Mustard

Quest. Alternaria blight of mustard is caused by

Ans. ***Alternaria brasicae***

Quest. White rust of crusifers is a

Ans. **Pseudo rust**

Quest. White rust/Blister is caused by

Ans. ***Albugo candida***

(7) Soybean

Quest. Yellow Mosaic of soybean is caused by

Ans. **Mungbean yellow mosaic virus (MYMV)**

Quest. The causal organism of Anthracnose/Pod blight

Ans. ***Colletotrichum truncatum***

(8) Sunflower and Sesame

Quest. Alternaria blight or leaf spot of sunflower is caused by

Ans. ***Alternaria helianthi***

Quest. Root and collar rot of sunflower is caused by

Ans. ***Sclerotium rolfssi***

Quest. The causal organism of Phyllody disease of sesame

Ans. **MLO**

(9) Cotton

Quest. Fusarium wilt is caused by

Ans. ***Fusarium moniliform***

Quest. The wilt of cotton is

Ans. ***Seed and Soil borne***

Quest. Black arm or bacterial blight is due to

Ans. ***Xanthomonas compestris***

Quest. Bacterial blight disease is

Ans. ***Internally seed borne***

Quest. Bacterial blight/Angular leaf spot is caused by

Ans. ***Xanthmonas malvacearum***

(10) Sugarcane

Quest. Most serious of sugarcane

Ans. **Red rot disease**

Quest. Red rot disease is caused by

Ans. ***Colletrotricum falcatum***

Quest. Red strip of sugarcane is caused by

Ans. ***Pseudomonas riubrilinus***

Quest. The causal organism of sugarcane smut

Ans. ***Ustilego citamini***

Quest. Grassy shoot of sugarcane is cased by

Ans. ***MLO***

Quest. The pith of the red rot affected can emits

Ans. **Rotten fish like smell**

Quest. The whip smut of sugarcane is caused by

Ans. ***Ustilago hordei***

(11) Potato and Tobacco

Quest. Most dangerous disease of potato

Ans. **Late blight**

Quest. Late blight of potato is caused by

Ans. ***Phytophthora infestans***

Quest. Early blight of potato is caused by

Ans. ***Alternaria solani***

Quest. The tuber borne disease of potato

Ans. **Black scurf (*Rhizoctonia solani*)**

Quest. Which potato disease causes Irish famine (1845)?

Ans. **Late blight**

Quest. Wart disease of potato is caused by

Ans. ***Synchutium endobioticum***

Quest. Potato virus diseases are spread by

Ans. **Aphids**

Quest. Damping off of tobacco is caused by

Ans. ***Pythium aphanidermaum***

Quest. Tobacco Mosaic disease is caused by

Ans. **Nicotiana Virus-1**

Quest. Root knot disease of tobacco is effective controlled by

Ans. **Carbofuran**

(12) Mango

Quest. Mango Malformation is caused by

Ans. ***Fusarium monilliformae***

Quest. Mango malformation is common in

Ans. **North-West India**

Quest. Black tip/Mango necrosis is caused by

Ans. ***Boron deficiency***

(13) Other Fruit Diseases

Quest. Scab disease in apple is caused by

Ans. ***Venturia inaequalis***

Quest. Anthracnose of guava is caused by

Ans. ***Collectotrichum psidii***

Quest. Leaf curl and mosaic of papaya is caused by

Ans. ***Virus***

Quest. Panama wilt of banana is also called as

Ans. ***Fusarium wilt***

Quest. Bunchy top of banana is caused by

Ans. ***Virus***

Quest. Which pathogen caused heavy losses to wine industry in France due to its epidemics in 1875?

Ans. ***Plasmopara viticola***

Quest. Citrus canker is caused by

Ans. ***Xanthomonas compestris pv citri***

Quest. Citrus gumosis is caused by

Ans. ***Phytophthora palmivora***

Quest. Mottle leaf of citrus is due to deficiency of

Ans. **Zinc**

Quest. Greening of citrus is caused by

Ans. ***Gracillicuts* (a *gram negative bacteria*)**

Quest. Downy mildew of grape vine is controlled by

Ans. **Bordeaux mixture**

(14) Tomato and Brinjal

Quest. Leaf curl of tomato is spread by

Ans. **White fly**

Quest. Early blight of tomato is caused by

Ans. ***Alternaria solani***

Quest. Wilting in Brinjal is caused by

Ans. ***Pseudomonas solanacearum***

Quest. Phomopsis blight or Fruit rot of Brinjal is caused by

Ans. ***Phomopsis vexans***

Quest. Damping off of Brinjal seedlings is due to

Ans. ***Pithium sp***

(15) Other Vegetable Diseases

Quest. Yellow vein mosaic of okra is transmitted through

Ans. ***White fly***

Quest. Damping off of chilly seedlings is due to

Ans. ***Pithium sp***

Quest. Anthracnose/Ripe rot/Die back of chilly is caused by

Ans. ***Collectotrichum capsici***

Quest. White blister of cabbage is caused by

Ans. ***Albugo candida***

Quest. Black rot of cabbage is caused by

Ans. ***Xanthomonas compestris***

Quest. A cabbage disease which is known to be more severe in acidic soils is

Ans. **Black rot**

Quest. Powdery mildew of cucurbits is caused by

Ans. ***Erysiphe cichoracearum***

Quest. Downey mildew of cucurbits is caused by

Ans. ***Pseudopernospora cubensis***

Quest. Downy mildew disease can be effectively managed by spraying of

Ans. **Metalaxyl**

Chapter 8
Horticulture

(I) Basic Horticulture

Quest. The term 'Horticulture' is derived from

Ans. **Latin word**

Quest. The science of growing of fruits, vegetables, ornament plants and preservation of foods

Ans. **Horticulture**

Quest. The science of production of fruit crops

Ans. **Pomology**

Quest. The term 'Pomology' is derived from

Ans. **Latin word**

Quest. The science of growing vegetable crops

Ans. **Olericulture**

Quest. The science of growing flower and ornamental plants.

Ans. **Floriculture/Ornamental horticulture**

Quest. A method by which food is kept out from spoilage after harvest

Ans. **Preservation**

Quest. Which process is involved in the senescence of fruits and vegetables?

Ans. **Respiration**

Quest. Fruits experiencing sudden increase in the rate of respiration at the time of ripening.

Ans. **Climacteric Fruits**

Quest. Mango, Banana, Guava, Papaya, Jackfruit, Sapota, Apple are known as

Ans. **Climacteric Fruits**

Quest. Fruits experiencing simple gradual decline in the rate of respiration at the time of ripening.

Ans. **Non-Climacteric Fruits**

Quest. Litchi, Lemon, Citrus, Grape, Ber, Pineapple are known as

Ans. **Non-Climacteric Fruits**

Quest. Which operation controls the shape of plant?

Ans. **Training**

Quest. Most widely used training system for commercial fruits

Ans. **Modified Leader System**

Quest. Removal of any excess or undesirable/unproductive branches, shoots or any other parts of plants

Ans. **Pruning**

Quest. Heading back and thinning out are associated with

Ans. **Pruning**

Quest. The most common pit size for fruit planting

Ans. **1 m × 1 m × 1 m**

Quest. The simplest system of fruit planting

Ans. **Square System**

Quest. In which planting system, a tree is planted on a corner of each angle

Ans. **Triangle System**

Quest. The 'filler tree technology' is associated with

Ans. **Quincunx System**

Quest. A pruning process in which a circular ring of bark measuring about 3 cm in length is removed.

Ans. **Ringing/Girdling**

Quest. Which fruit has the highest Vitamin A?

Ans. **Mango**

Quest. Which grafting is used for repairing the plant?

Ans. **Bridge grafting**

(II) CULTIVATION OF IMPORTANT FRUIT CROPS

(1) Mango

Quest. Mango is also known as

Ans. **King of fruits/National fruit/ Bathroom fruit**

Quest. The botanical name of mango

Ans. ***Mangifera indica***

Quest. Mango belongs to which family

Ans. **Anacardiaceae**

Quest. The origin place of mango

Ans. **Indo-Burma region**

Quest. Leading Mango producing state having maximum area under mango

Ans. **Uttar pradesh**

Quest. The commercial propagation method of mango

Ans. **Veneer grafting**

Quest. The normal planting space of mango

Ans. **10m × 10m**

Quest. High density planting (2.5 m × 2.5 m) of mango is done in

Ans. **Amrapalli variety**

Quest. Most popular variety of India

Ans. **Alphanso**

Quest. Sweetest variety of mango

Ans. **Chousa**

Quest. Regular bearer varieties of mango

Ans. **Ratna, Neelum, Himsagar, Gulab khas, Pairy and Totapari**

Quest. Seedless variety of mango

Ans. **Sindhu**

Quest. Mallika is a cross of

Ans. **Neelam × Dashehari**

Quest. Amrapalli is a cross of

Ans. **Dashehari × Neelam**

Quest. Ratna is a cross of

Ans. **Neelam × Alphanso**

Quest. Sindhu is a cross of

Ans. **Ratna × Alphanso**

Quest. Most commonly used for flower induction

Ans. **Paclobutrazol**

Quest. The pollinator of mango is

Ans. **Housefly**

Quest. Bearing habit of mango is

Ans. **Terminal**

Quest. Fruit drop in mango is controlled by

Ans. **2, 4-D**

Quest. Mango malformation is controlled by

Ans. **NAA**

Quest. Spongy tissue is due to

Ans. **Convection heats**

Quest. Internal fruit necrosis is due to

Ans. **Boron deficiency**

(2) Guava

Quest. The botanical name of guava

Ans. ***Psidium guajava***

Quest. The normal planting space of guava

Ans. **10m × 10m**

Quest. The commercial propagation method of guava

Ans. **Air layering**

Quest. Which guava variety is known as Sardar

Ans. **Lucknow- 49**

Quest. Dual purpose variety of guava

Ans. **Lalit**

Quest. Parthenocarpic variety of guava

Ans. **Allahabad round**

Quest. Kohir safed is a cross of

Ans. **Kohir × Allahabad Safeda**

Quest. Safed Jam is a cross of

Ans. **AS × Kohir**

Quest. The fruiting time of Mrig bahar

Ans. **Nov-January**

(3) Papaya

Quest. Origin of Papaya

Ans. **Tropical America**

Quest. Yellow pigment in papaya

Ans. **Caricaxanthin**

Quest. The commercial propagation method of papaya

Ans. **Seed (500 g/ha)**

Quest. Planting space of papaya

Ans. **2m×2m**

Quest. Best suited variety for high density planting of papaya

Ans. **Pusa nanha (1.25m×1.25m)**

Quest. Highest papain yielding variety

Ans. **Pusa majesty**

Quest. Pusa delicious, Pusa majesty, CO-3 and Coorg honew dew are

Ans. **Gynodioecious varieties**

Quest. The serious disease of papaya

Ans. **Damping off**

Quest. The fruiting time of papaya

Ans. **Feb-June**

Quest. The chemical used for better colour and keeping quality of papain

Ans. **Potassium meta-bi-sulphite (KMS)**

Quest. Enzyme present in dried latex of papaya

Ans. **Pepsin**

Quest. Pusa dwarf, Pusa Nanha and Pusa giant are naturally

Ans. **Didecious**

(4) Pomegranate

Quest. The botanical name of pomegranate

Ans. ***Punica granatum***

Quest. The normal planting space is

Ans. **6m × 6m**

Quest. Commercial propagation method

Ans. **Air layering**

Quest. The hybrid variety of pomegranate

Ans. **Amlidana**

Quest. Most popular vary

Ans. **Ganesh and Dholka**

Quest. Fruit cracking is most probable in

Ans. **Mrig bahar season**

(5) Citrus

Quest. The botanical name of acid lime

Ans. ***Citrus aurentifolia***

Quest. Break fast fruit is

Ans. **Grape fruit (*C. paradise leaf*)**

Quest. The botanical name of Sweet Orange

Ans. ***Citrus sinensis***

Quest. Which citrus species is known as *Fancy fruit* ?

Ans. ***C. reticulate***

Quest. Thornless species of citrus

Ans. **Tahiti lime (*C. latifolia*)**

Quest. Monoembryonic species of citrus

Ans. **Pumelo**

Quest. Polyembryonic species of citrus

Ans. **Acid lime**

Quest. Which species of citrus produces seedless fruits?

Ans. **Tahiti lime**

Quest. The normal planting space of citrus

Ans. **5-6m × 5-6m**

Quest. Commercial propagation method

Ans. **Seed and Budding**

Quest. Rootstock of mandarin orange is

Ans. **Rangpur lime**

Quest. Best method for irrigation of citrus

Ans. **Ring method**

Quest. 'Kinnow' is a cross between

Ans. **King × Willow leaf**

Quest. Seedless variety of mandarin orange

Ans. **Satsuma**

Quest. Lucknow seedless is a variety of

Ans. **Lemon**

(6) Banana

Quest. The banana is commonly known as

Ans. **Adam's fig and Tree of paradise**

Quest. The normal planting space of banana

Ans. **1.8-2m × 1.8-2m**

Quest. Commercial propagation method

Ans. **Sword Suckers**

Quest. Banana inflorescence is known as

Ans. **Spadix**

Quest. Best variety for chips making

Ans. **Narendran**

Quest. Gold finger is a

Ans. **Hybrid banana**

Quest. How much sugar contains in ripe banana?

Ans. **26%**

Quest. Degreening of banana is done by

Ans. **Ethylene**

Quest. Removal of male bud after completion of female phase is known as

Ans. **Denavelling**

Quest. Removal of undesired suckers, done once in 45 days of planting

Ans. **Desuckering**

Quest. *Tetrazolium test* is used for detection of

Ans. **Bunchy top virus**

(7) Other Fruit Crops

Quest. Which crop is commonly known as *Single seeded nut?*

Ans. **Litchi**

Quest. Litchi is commercial propagated by

Ans. **Air layering and Seed**

Quest. Red pigment in litchi is due to

Ans. **Anthocyanin**

Quest. Sapota is grown in

Ans. **Tropical climate**

Quest. Commercial propagation method of sapota

Ans. **Inarching**

Quest. Most popular varieties of sapota

Ans. **Kali patti, Pili patti, Oval, Cricket ball, Chatri, Barahmasi.**

Quest. CO-1 is a cross of

Ans. **Cricket ball × oval**

Quest. Botanical name of Monkey jack

Ans. ***Autocarpus heterophylus***

Quest. Jackfruit is commercially propagated by

Ans. **Air layering**

Quest. The popular varieties of jackfruit

Ans. **Champa, Rudrakshi and Singapore**

Quest. Aonla is commercially propagated by

Ans. **Inarching**

Quest. The popular varieties of Aonla

Ans. **Banarasi, Krishna, Chakaiya, Hathi jhul, Kanchan, NA-7, 9**

Quest. Most widely used training system of aonla

Ans. **Modified central leader system**

Quest. Ber is commonly known as

Ans. **Poor man's fruit and king of arid fruits**

Quest. The commercially cultivated variety of ber

Ans. **Umran**

Quest. Early variety of ber

Ans. **Seb**

Quest. Best time for prunning of ber

Ans. **End of May to Mid June**

Quest. Ber fruits are matured at

Ans. **5-6 months after flowering**

Quest. Richest source of Vit-B_2

Ans. **Bael**

Quest. Active ingredient present in Bael

Ans. **Marmelosin**

Quest. The most ideal stage of bael for making preserve

Ans. **Mature green stage**

Quest. Paras is a variety of

Ans. **Jamun**

Quest. Seedless variety of Jamun is

Ans. **Narendra Jamun-6**

Quest. The commercial propagation method of pineapple

Ans. **Suckers and slips**

Quest. Singapore, Mauritious and Giant Kew are the varieties of

Ans. **Pineapple**

Quest. The enzyme contains in pineapple fruit

Ans. **Bromelin**

Quest. The chemical used for inducing flowering in pineapple

Ans. **Ethrel and NAA**

Quest. Queen of temperate fruit

Ans. **Apple**

Quest. Apple bowl of India

Ans. **Himachal Pradesh**

Quest. Apple is commercially propagated by

Ans. **Tongue grafting and Whip budding**

Quest. Redness in apple is due to

Ans. **Anthocyanin**

Quest. Delicious, Rome beauty and Parlin's Beauty are

Ans. **Late maturing apple**

Quest. Diploid variety of apples are

Ans. **Self fertile**

Quest. Usually apple is graded on

Ans. **6 size**

Quest. Discolouration of apple after cutting is due to

Ans. **Enzymes**

Quest. Commercial propagating method of grapevine

Ans. **Hard wood cutting**

Quest. Arka Hans is a cross of

Ans. **Banglore Blue × Anab-e-shahi**

Quest. Most widely accepted training system of grapevine in India

Ans. **Bower system**

Quest. Which is used for improving fruit quality of grape?

Ans. **20 ppm GA**

Quest. Which fruit is commonly known as "*Kalpavriksha*

Ans. **Coconut**

Quest. Cashew nut and almond are the richest source of

Ans. **Fat**

Quest. Richest source of Iron

Ans. **Karonda**

(III) CULTIVATION OF IMPORTANT VEGETABLE CROPS

(1) Tomato

Quest. No. 1 processing vegetable

Ans. **Tomato**

Quest. The new botanical name of tomato

Ans. ***Solanum lycopersicon***

Quest. A pigment responsible for red colour in tomato

Ans. **Lycopene**

Quest. Pusa Rubi is a cross of

Ans. **Sioux × Improved Maruti**

Quest. The best combiner variety of tomato

Ans. **Pusa Rubi**

Quest. Pusa Rubi, Arka sourav, Pant bahar and Best of all are

Ans. **Indeterminate varieties**

Quest. The best suited variety for drought condition

Ans. **Arka Vikas**

Quest. Most serious pest for tomato plant

Ans. **Root knot nematode**

Quest. Nematode and Bacterial wilt resistant variety of tomato

Ans. **Arka vardan**

Quest. Most important nutrients required for tomato cultivation

Ans. **Boron and zinc**

Quest. The recommended seed rate of tomato per hectare

Ans. **300-350 gm/ha**

Quest. Harvesting stage of tomato for distant market/transportation

Ans. **Mature green stage**

Quest. Best method of extraction of tomato seed

Ans. **Alkali method**

Quest. Blossom end rot of tomato is due to

Ans. **Ca deficiency**

Quest. Fruit cracking of tomato is due to

Ans. **B deficiency**

(2) Brinjal

Quest. Brinjal is also known as

Ans. **Egg plant**

Quest. Brinjal fruits are good source of

Ans. **Vit.-B**

Quest. White brinjal is preferred by

Ans. **Diabetics patients**

Quest. The normal seed rate of brinjal

Ans. **200 g/ha**

Quest. Extra early maturing variety of brinjal

Ans. **Pusa purple long**

Quest. Phomopsis blight and Bacterial blight resistant variety of brinjal

Ans. **Pant Samrat**

Quest. The normal seed rate of brinjal

Ans. **100 sq.m/ha**

Quest. Nursery area of brinjal

Ans. **1.5-2.0 kg/ha**

(3) Chilly

Quest. The botanical name of chilly

Ans. ***Capsicum annum***

Quest. Causes of red colour in chilli

Ans. **Capsenthin**

Quest. The cause of pungency in chilli

Ans. **Capsicin**

Quest. Variety suitable for HDP

Ans. **Jwalamukhi**

Quest. Leaf curl resistant varieties

Ans. **Pusa Jwala, Pusa Sadabahar, Pant C-1**

Quest. The chemical used for fruit setting

Ans. **Triacontanol**

Quest. Green to dry chilli ratio

Ans. **10:1**

(4) Cucurbits

Quest. Cultivation practise followed in cucurbits

Ans. **Daria cultivation**

Quest. Cultivated pumpkin is botanically known as

Ans. ***Cucurbita moschata***

Quest. Seed rate of pumpkin

Ans. **1.0-1.5 kg/ha**

Quest. Chief pollinator of pumpkin

Ans. **Honey bee**

Quest. PKM 1 is a variety of

Ans. **Snake guord**

Quest. Botanical name of bottle gourd

Ans. ***Lagenaria siceraria***

Quest. Seed rate of bottle gourd

Ans. **3-4 kg/ha**

Quest. Popular varieties of bottle guard

Ans. ***Pusa Summer Prolific long, Summer Prolific Round, Pusa Manjari, Pusa Megdoot***

Quest. Which cucurbit is also known as Bitter cucumber

Ans. **Bitter gourd**

Quest. Seed rate of bitter gourd

Ans. **4.5-5 kg/ha**

Quest. Gynomonoecious flowers are found in

Ans. **Cucumber**

Quest. Pusa Sanyog is a variety of

Ans. **Cucumber**

Quest. Fruit type of cucumber is

Ans. **Pepo**

Quest. Bitterness in cucumber is due to

Ans. **Metaxenia**

Quest. Pusa Nasdar and Satputia are popular varieties of

Ans. **Ridge guard**

Quest. *Trichosanthus dioca* is botanical name of

Ans. **Pointed gourd**

Quest. Pointed gourd is propagated through

Ans. **Vine cutting**

Quest. Pusa Chikni, Pusa Supriya and Harita are the varieties of

Ans. **Sponge guard**

Quest. How much water contains by water melon fruit?

Ans. **95%**

Quest. The seed rate of water melon

Ans. **3.5-5 kg/ha**

Quest. Sugar Baby is the popular variety of

Ans. **Water melon**

Quest. Pusa bedana is a cross of

Ans. **Tetra-2 × Pusa Rasal**

Quest. Pink bedana is the variety of

Ans. **Musk melon**

Quest. Fruit of okra is known as

Ans. **Capsule**

(5) Other Vegetables

Quest. Richest source of protein is

Ans. **Beans**

Quest. The seed rate of okra is

Ans. **8-10 kg/ha**

Quest. Pusa Sawani and Parbhani Kranti is famous variety of

Ans. **Okra**

Quest. Yellow vein mosaic resistant variety of okra

Ans. **Pusa Sawani**

Quest. A deep rooted crop is

Ans. **Sweet potato**

Quest. Edible part of cabbage

Ans. **Head**

Quest. Seed rate of cabbage is

Ans. **350-500 gm/ha**

Quest. Anti-cancer property of cabbage is due to

Ans. **Indole-3-Cardinal**

Quest. The botanical name of cauliflower

Ans. ***Brassica oleracia.var botrytis***

Quest. Seed rate of cauliflower is

Ans. **500-600 gm/ha**

Quest. Edible part of cauliflower

Ans. **Curd**

Quest. The important process of cauliflower

Ans. **Blanching**

Quest. Pusa Snowball is a variety of

Ans. **Cauliflower**

Quest. Seed rate of knol khol is

Ans. **1-1.5 kg/ha**

Quest. Economic part of knol khol is

Ans. **Extended stem**

Quest. Temperature required for bolting in onion is

Ans. **< 15 ^{0}C**

Quest. Japanese white is a variety of

Ans. **Raddish**

Quest. The variety of carrot which is richest source of vitamin A

Ans. **Pusa Meghali**

Quest. IIHR is located at

Ans. **Bangalore**

Quest. Which growth regulator is isolated from yam?

Ans. **Batasin**

Quest. Only tuber crop, which is rich in protein?

Ans. **Colocasia**

Quest. Toxic substance present in colocasia

Ans. **Ca oxalate**

Quest. The product of cassava is

Ans. **Sago**

Quest. Pungency in garlic is due to

Ans. ***Alicin, Allinase***

Quest. Economical part of sweet potato

Ans. **Adventious roots**

Quest. Pungency of onion is due to

Ans. ***Allyl Propide di-sulphide***

Quest. Irritation of eye due to cutting onion is due to presence of

Ans. **Pyruvic acid**

(IV) ORNAMENTAL HORTICULTURE

Quest. Leading cut flower exporter in the world is

Ans. **Netherlands**

Quest. Concept of lawn was developed in

Ans. **England**

Quest. In which garden, arrangement of rocks is main feature?

Ans. **Japanese garden**

Quest. Shrubs or trees planted at regular intervals on boundry for fencing

Ans. **Hedges**

Quest. The green carpet for the landscape maintained by growing and mowing grasses.

Ans. **Lawn**

Quest. Thorny fencing plant used as a hedge

Ans. ***Inga dulcus***

Quest. Planting of low growing plants along with paths, roads, flower beds, lawns etc. for demarcation and beautification

Ans. **Edge**

Quest. The art of developing the plant or training the plant into different forms or shapes like animals, birds, arches, etc.

Ans. **Topiary**

Quest. The arrangement of colourful potted plants in different tiers around a central object which may be tree trunk, lamp post or pillar.

Ans. **Trophy**

Quest. Growing of shrubs in a group

Ans. **Shrubbery**

Quest. A group of ornamental plants used to grow over walls, trellis, arches, pergolas, arbours, pillars, bowers *etc.*

Ans. **Climbers and Creepers**

Quest. An art of growing and training of a plant to a miniature form having a natural look of old age.

Ans. **Bonsai**

Quest. The father of rose breeding

Ans. **Dr. Bhattachaterji**

Quest. The fruit of rose is known as

Ans. **Hip**

Quest. Rose can be cultivated up to

Ans. **5 years**

Quest. Commercial propagation method of rose

Ans. **'T' budding**

Quest. Floribundas is a cross of

Ans. **Hybrid tea × Dwarf polyantha**

Quest. Yellow coloured rose species is

Ans. ***R. foitida***

Quest. Thornless variety of rose

Ans. **Chitra**

Quest. The single borne rose species

Ans. **Hybrid tea**

Quest. Crimson Glory and Super Star are which type of rose

Ans. **Hybrid tea**

Quest. Rose species having large flowers in clusters

Ans. **Floribundas**

Quest. The growth habit of Chrysanthemum

Ans. **Perennial**

Quest. Commercial propagation method of Chrysanthemum

Ans. **Root suckers and Terminal cuttings**

Quest. Which chrysanthemum allows single bloom on a branch?

Ans. **Standard Chrysanthemum**

Quest. Off-season variety of chrysanthemum

Ans. **Haldi ghati, Himansu, Jaya and Jwala**

Quest. Per hectare of seed rate of marigold

Ans. **1.2-1.5 kg/ha**

Quest. Commercial propagation method of gladiolus

Ans. **Corms**

Quest. Annual carnation is propagated by

Ans. **Seeds**

Quest. True marigold is also known as

Ans. **Calendulas**

Quest. Golden Age and Crown of Gold are the variety of

Ans. **African Marigold**

Quest. Tuber is commercially propagated by

Ans. **Tubers**

Quest. Queen Elizabeth is a variety of

Ans. **Floribundas rose**

Quest. The botanical name of hollyhock

Ans. ***Althaea rosea***

Quest. The flower colour of Chrysanthemum

Ans. **Yellow and white**

(V) POST HARVEST TECHNOLOGY

Quest. Who is known as Father of food preservation?

Ans. **Nicolas Apart**

Quest. A heat treatment food material at 72^{0}C for 15 seconds, or 63^{0}C for 30 minutes

Ans. **Pasteurization**

Quest. Juices are mostly preserved by

Ans. **Freezing**

Quest. The original colour of beverages for longer period are retained by

Ans. **Benzoic acid**

Quest. Removal of moisture from the food materials for preservation

Ans. **Dehydration**

Quest. Which of the following is used for killing microorganisms in food?

Ans. **Heat processing**

Quest. The concentration of sugar required for preservation of fruits and jam

Ans. **66%**

Quest. The concentration of salt sufficient to preserve most of the food products

Ans. **15-25%**

Quest. A thermal process mostly used for vegetables prior to freezing, drying, or canning in order to soften the texture.

Ans. **Blanching**

Quest. The peeling of fruits and vegetables is known as

Ans. **Lye peeling**

Quest. The solution made by dissolving Salt (NaOH) in water is called as

Ans. **Brine solution**

Quest. Which fruit beverage contains atleast 10% fruit juice and 10% soluble solids?

Ans. **Ready-To-Serve (RTS)**

Quest. The fruit beverage commonly prepared from Mango, Papaya, Bel, Aonla

Ans. **Nectar**

Quest. A fruit juice normally contains 25% juice and 40% TSS

Ans. **Squash**

Quest. Mango pulp is preserved by

Ans. **Sugar**

Quest. Which fruit beverage is diluted before serving?

Ans. **Syrup**

Quest. TSS of jam should not be

Ans. **< 70%**

Quest. TSS of cooking jam is measured by

Ans. **Hand Refractometer**

Quest. A semi-solid transparent product prepared from pectin containing fruit

Ans. **Jelly**

Quest. For jelly making, fruit should be harvested at

Ans. **Firm ripe stage**

Quest. The pH of final jelly should be

Ans. **3.2**

Quest. The instrument used to know pectin content

Ans. **Jellimeter**

Quest. A fruit or vegetable impregnated with the cane sugar or glucose syrup

Ans. **Candied fruit or vegetable**

Quest. A product prepared from strained pulp of fully ripe tomato fruits after cooking

Ans. **Tomato Sauce/Ketchup**

Quest. TSS% of Tomato sauce is

Ans. **30%**

Quest. The salt concentration in pickle is maintained at

Ans. **8-10%**

Quest. CFTRI denotes

Ans. **Central Food Technological Research Institute, Mysore**

Chapter 9
Agricultural Extension and Economics

(I) Agricultural Extension

Quest. An educational process to bring about desirable changes

Ans. **Extension**

Quest. 'Extension' is a

Ans. **Latin word**

Quest. Extension activity was started first time in

Ans. **USA**

Quest. The term "Extension education" was originated from

Ans. **England (1866)**

Quest. The term "Extension education" was firsed used by

Ans. **Cambridge University, England (1873)**

Quest. The father of extension education in India

Ans. **J.P. Leagans**

Quest. Extension education is both

Ans. **Science and Art**

Quest. The basic principle of extension education is

Ans. **Help to those who helps themselves.**

Quest. The right approach of Agril. Extension

Ans. **Bottom up approach**

Quest. 'EDUCARE' (Latin word) means

Ans. **To bringup physically or mentally.**

Quest. A statement of situation, objectives, problems and solutions

Ans. **Programme**

Quest. The process by which human behavior is modified

Ans. **Education**

Quest. Extension education is

Ans. **Informal educaion**

Quest. School education is

Ans. **Formal education**

Quest. Situation, Objectives, teaching, evaluation and reconsideration are the steps of

Ans. **Extension education**

Quest. Attention → interest → desire → conviction → action → satisfaction are the steps of

Ans. **Extension teaching**

Quest. An understood information possessed by a person

Ans. **Knowledge**

Quest. The process of working with rural people in an effort to recognize the problems and determine possible solutions.

Ans. **Programme planning**

Quest. The father of rural sociology

Ans. **August Compte**

Quest. The science of human behaviours

Ans. **Psychology**

Quest. The interchange of ideas between two persons, in such a way that they act on the existing knowledge to achieve some useful results

Ans. **Communication**

Quest. The suitable medium to establish commonness between sender and receiver of message

Ans. **Communication**

Quest. Communication is a

Ans. **Two way/Double way Process**

Quest. Shannon and Weaver (1949) proposed model of communication

Ans. **Source → Transmitter → Signal → Receiver → Destination**

Quest. Berlo Model of communication (1960)

Ans. **Source → Encoder → Message → Signal → Decoder → Receiver**

Quest. Leagans Model (1963) is

Ans. **Communicator → Message → Channel → Treatment → Audience → Audience response**

Quest. Speaker → Speech → Audience is a communication model proposed by

Ans. **Aristotle**

Quest. A specific way adopted by the communicator to communicate his message effectively so that whole message is understood by maximum number of audience.

Ans. **Treatment of message**

Quest. The process by which an innovation is communicated through certain channels overtime among the members of a social system

Ans. **Diffusion**

Quest. A decision to continue full use of an innovation

Ans. **Adoption**

Quest. The mental process through which individual passes from fires hearing about an innovation to final adoption.

Ans. **Adoption process**

Quest. Stages of adoption are

Ans. **Awareness → Interest → Evaluation → Trial → Adoption**

Quest. An attention with a sense of concerns focused upon some object

Ans. **Interest**

Quest. Which stage of adoption helps to establish "Bench mark"?

Ans. **Evaluation**

Quest. The people who adopt immediately after getting knowledge and constitutes only 2.5% of the total population.

Ans. **Innovators**

Quest. The people adopt through local leaders and constitute only 13.5% of the total population.

Ans. **Early adopters**

Quest. The percentage population of early majority over total population

Ans. **34%**

Quest. The farmer who accepts new practices very last with in his social system is known as

Ans. **Laggard (16 per cent)**

Quest. A process by which an idea or innovation spreads

Ans. **Diffusion**

Quest. The process of arranging situations that stimulate and guide learning activities in order to bring desirable changes in the behaviour of people

Ans. **Teaching**

Quest. Traditional teaching method

Ans. **Drama**

Quest. Central element in learning situation

Ans. **Learner**

Quest. A mental and/or physical reaction that makes through seeing

Ans. **Learning experience**

Quest. Formulae of Intelligent Quotient

Ans. **I.Q. (%) =**

$$\frac{\textbf{Mental age}}{\textbf{Chronological age}} \times 100$$

Quest. The method of face-to-face or person-to-person contact between the rural people and extension workers

Ans. **Individual contact**

Quest. Farm and home visit, Office calls, Telephone calls, Personal letters, Result Demonstration are comes under

Ans. **Individual contact**

Quest. A method in which 20 to 30 rural people or farmers are contacted in a group

Ans. **Group-contact**

Quest. Conferences, Pannel, Symposium, Discussion, Meeting, Workshops, Field trips, Tour are comes under

Ans. **Group contact**

Quest. The media used for mass contact

Ans. **Radio, Television, Exhibitions, Bulletins, Leaflets, News letter, Circular letters, Posters, folder/pamphlet etc.**

Quest. A sheet of paper with pictorial slogan, which is utilized to attract the mass attention for single idea.

Ans. **Poster**

Quest. The most common size of poster

Ans. **50cm x 75cm**

Quest. Round table discussion is called

Ans. **Panel**

Quest. Mostly widely used pamphlet size

Ans. **12 to 24 pages**

Quest. A published material on a small paper in which there is brief information of a subject

Ans. **Leaflet**

Quest. When a paper folds ones or twice with detail information on specific aspects is called as

Ans. **Folder**

Quest. A small published book consisting of 24 to 48 pages

Ans. **Bulletin**

Quest. A series of illustrated cards flashed before a group in proper sequence to tell a complete story step by step to the group of the learners.

Ans. **Flash cards**

Quest. For a group of 10-25, people use flash cards of size

Ans. **10"x12" (25 cm x 30 cm)**

Quest. The letter used to send the same information to many people at the same time is called

Ans. **Circular Letter**

Quest. When two or more brief talks presenting phases of the some general topic called

Ans. **Symposium**

Quest. A systematic display of models, specimens, charts, real objects and any informative materials.

Ans. **Exhibition**

Quest. The basic principle of Demonstration

Ans. **Learning by seeing and doing**

Quest. The concept of demonstration was given by

Ans. **Dr. Seeman A. Knapp**

Quest. Method demonstration is

Ans. **Short -type demonstration**

Quest. The oldest form of teaching

Ans. **Method demonstration**

Quest. The main purpose of method demonstration is

Ans. **To provide skill**

Quest. The basic principle of method demonstration

Ans. **Learning by doing**

Quest. The father of method demonstration

Ans. **Dr. Seeman A. Knapp**

Quest. A single practice demonstration used to show method of sowing

Ans. **Method demonstration**

Quest. A demonstration practice used to compare two technologies *i.e.* old and new

Ans. **Result demonstration**

Quest. The demonstration used to improve skill, knowledge and attitude

Ans. **Result demonstration**

Quest. Result demonstration is based on

Ans. **Seeing by doing**

Quest. The front-line demonstrations conducted by researchers on the farmers field

Ans. **National demonstrations**

Quest. Tape-recorder, Radio and Telephone are

Ans. **Audio Aids**

Quest. Non-projected visual aids

Ans. **Posters, Charts, Flashcards, Bulletin board, Photograph etc.**

Quest. Television is a type of

Ans. **Projected Audio Visual Aid**

Quest. The best media to communicate with farmers and for village people

Ans. **Puppets**

Quest. A transparent picture or photograph in an individual mount, projected through slide projector

Ans. **Slides**

Quest. The basic unit of civilization

Ans. **Family**

Quest. The basic unit of rural society

Ans. **Village**

Quest. A family consists of husband, wife and their children known as

Ans. **Nuclear family**

Quest. The uniformly accepted ways of acting about some social aspects of life are known as :

Ans. **Custom**

Quest. They are uniformly accepted ways of thinking

Ans. **Tradition**

Quest. Cooperative Movement (1904) was initiated by

Ans. **F. Nicholson**

Quest. Concept of Village Level Worker was related with the programme

Ans. **Sri Niketan**

Quest. Gurgaon Project (1920) was started by

Ans. **Mr. F.L. Brayne**

Quest. Young Men Christian Association was associated with

Ans. **Marthandom Project (1928)**

Quest. Rural Development programme was started on

Ans. **1935**

Quest. The district level extension programme launched in independent India was in

Ans. **Etawah**

Quest. Etawah Pilot Project (1948) was initiated by

Ans. **Albert Mayer**

Quest. Five year plans were started on

Ans. **1951**

Quest. Community Development Project (CDP) was started on

Ans. **2nd Oct, 1952**

Quest. Which programme is called as Package programme

Ans. **Intensive Agriculture District Programme (1960)**

Quest. High Yielding Varietal Programme (HYVP) was started on

Ans. **1966-67**

Quest. The programme initiated at occasion of ICAR Golden Jubille celebration

Ans. **Lab to Land Programme (1979)**

Quest. Training and Visit programme (1974) is also known as

Ans. **Baster and Benor Scheme**

Quest. Who had recommended Panchayat Raj System?

Ans. **Balwant Rai Mehata Committee**

Quest. The basic principle/slogan of TRYSEM was

Ans. **Learning by doing**

Quest. The primary aim of Integrated Rural development Programme (IRDP)

Ans. **All round development of family**

Quest. NABARD is started on

Ans. **12th July, 1982**

Quest. A programme to provide atleast 100 days wage employment in rural areas

Ans. **National Rural Employment Guarantee Act (NAREGA, 2006)**

Quest. A guarantee programme for the people below poverty line.

Ans. **National Food Security Mission (2007)**

(II) Agricultural Economics

Quest. The science of Wealth

Ans. **Economics**

Quest. The economic concerned with individual unit *i.e.,* single industry, form or single consumer.

Ans. **Micro economics**

Quest. The economic deal with the whole economic setup *i.e.* total production, total expenditure, total income etc.

Ans. **Macro economics**

Quest. Father of Agricultural Economics

Ans. **Adam Smith**

Quest. The potential exchangeable means of satisfying human wants

Ans. **Wealth**

Quest. The part of wealth used for further regenerating wealth

Ans. **Capital**

Quest. Reward of Labour

Ans. **Wages**

Quest. Reward of Capital

Ans. **Interest**

Quest. Population theory was proposed by

Ans. **Malthus**

Quest. The market used for food grains

Ans. **Regional or State market**

Quest. The market used for durable goods

Ans. **National market**

Quest. A time based market basically for perishable goods

Ans. **Short period market**

Quest. Food grain markets, vegetable markets, wool market are the example of

Ans. **Special market**

Quest. Which market ensure fair price

Ans. **Regulate market**

Quest. Which one is a competitive market?

Ans. **Perfect market**

Quest. The market which is permanent in nature

Ans. **Secular market**

Quest. The 1st function performed in the marketing of agricultural commodities

Ans. **Packing**

Quest. Which is not a function of marketing

Ans. **Harvesting**

Quest. FCI was established in

Ans. **1965**

Quest. The Warehousing Corporations Act came into operation on

Ans. **18th March, 1962**

Quest. NAFED was established in the year

Ans. **October, 1958**

Quest. When there is a single salers of a product

Ans. **Monopoly market**

Quest. A market consisting of single buyer of a product

Ans. **Monosony market**

Quest. When few salers of a commodity

Ans. **Oligopoly market**

Quest. When few buyers of a commodity

Ans. **Oligopsony market**

Quest. A market where homogenous products and large no. of buyers and salers are found

Ans. **Pure Market**

Quest. The contribution of central govt. in the capital share of Regional Rural Bank

Ans. **50%**

Quest. WTO come in to effect from

Ans. **1995**

Quest. The portion of the total produce in stock which the farmer is willing to sell.

Ans. **Marketable surplus**

Quest. The portion which is usually brought to the market at a particular time for sale.

Ans. **Marketed surplus**

Quest. The relationship between Marketable to Marketed surplus for perishable products

Ans. **Equal**

Quest. Marketable surplus is given by

Ans. **MS = P – C**

Quest. The minimum price at which the govt. is prepared to buy agril. commodities

Ans. **Minimum Support Price (MSP)**

Quest. MSP is fixed by

Ans. **Commission of Agricultural Cost and Price (CACP)**

Quest. National institute of Agricultural Marketing is situated at

Ans. **Jaipur**

Quest. The difference between value in use and value in exchange

Ans. **Consumer surplus**

Quest. The excess of what we are prepared to pay over what we actually pay for a commodity is known as

Ans. **Consumer surplus**

Quest. Short term loan is given for

Ans. **1 to 1½ years**

Quest. Long term loan is given for the period of

Ans. **5 to 30 years**

Quest. The 3 Rs of credit are

Ans. **Returns, Repayment capacity and Risk bearing ability.**

Quest. Loan repaying capacity of a farmer is judged on the basis of

Ans. **Net income of farmer**

Quest. Increase in money supply and fall in production causes

Ans. **Inflation**

Quest. Risks arise due to changes in Government policies are termed as a

Ans. **Institutional risk**

Quest. AGMARK Act was passed in

Ans. **1937**

Quest. Central AGMARK Lab is located at

Ans. **Nagpur**

Quest. 1st bank in India was established on

Ans. **1806**

Quest. Money supply in Indian national economy is regulated by

Ans. **RBI**

Quest. RBI was established and nationalized in

Ans. **1st April, 1935 and 1st January, 1949**

Quest. The bank who credits to marginal, small and Agricultural labours

Ans. **RRB (1975)**

Quest. The full form of NABARD

Ans. **National Bank for Agricultural and Rural Development**

Quest. 14 commercial banks were nationalization on

Ans. **19th July, 1969**

Quest. Income Tax is an example of

Ans. **Direct tax**

Quest. The tax which is levied on goods or services produced or purchased.

Ans. **Indirect tax**

Quest. A multistage sales tax with credit for taxes paid on business purchases.

Ans. **VAT**

Quest. A farmer having an area of <1 ha

Ans. **Marginal farmer**

Quest. The operational land holding of small farmers

Ans. **1-2 ha**

Quest. The land holding of large farmers

Ans. **> 10 ha**

Quest. On which basis, CACP fixed minimum prices of crops

Ans. **Cost of production**

Quest. A science of decision making

Ans. **Farm Management**

Quest. Production function is also known as

Ans. **Input Output relation**

Quest. Transformation of physical inputs into physical outputs is termed as

Ans. **Production**

Quest. Cost calculated per hectare is known as

Ans. **Cost of Production (COC)**

Quest. The most important unit of farm management

Ans. **Production unit**

Quest. The basis of Cobb - Douglas Production function is

Ans. **Constant elasticity of substitution**

Quest. When $MPP_X > APP_X$; then APP_X is

Ans. **Increasing**

Quest. Under perfect competition market, maximum profit is obtained when

Ans. **Marginal Return = Marginal Cost**

Quest. Optimum profit will be obtained at a point where

Ans. **MC = MP**

Quest. Marginal cost is equal to

Ans. $$\frac{\textbf{Change in total cost}}{\textbf{Change in output}}$$

Quest. When the demand and price are equal, called as

Ans. **Equilibrium price**

Quest. The principle applied for production function "How much to produce"

Ans. **Principal of diminishing returns/ costs**

Quest. The principle applied for "How to produce"

Ans. **Principal of least cost combination**

Quest. The principle applied for "What to produce"

Ans. **Principal of opportunity cost/ Equimarginal returns**

Quest. Optimum level of input use without resource limitation

Ans. **Law of diminishing return**

Quest. Basic fundamental law of agriculture is

Ans. **Law of diminishing return**

Quest. Choose best crop enterprises

Ans. **Principal of opportunity cost**

Quest. When total Assets are divided by total liabilities, called as

Ans. **Net Capital Ratio (NCR)**

Quest. The formulae of Rate of Turn Over

Ans. **Rate of Turn Over =**

$$\frac{\textbf{Gross Income}}{\textbf{Total Farm Assets}} \times 100$$

Quest. When MP = 0, then $E_P = 0$ is called

Ans. **Completely inelastic demand**

Quest. When MP > AP then $E_P > 1$ is called

Ans. **Elastic demand**

Quest. When MP = AP, then $E_P = 1$

Ans. **Unit inelastic demand**

Quest. Demand of Agri. Products are always comes under

Ans. **Unit inelastic demand**

Quest. Inflexion point is found on

Ans. **Irrational zone (Stage – I)**

Quest. The zone at which TP increases but at decreasing rate

Ans. **Rational zone (Stage – II)**

Quest. EP is always less than zero in

Ans. **Irrational zone (Stage – III)**

Quest. Price Ratio is equal to

Ans. **Cost per unit of added resource**

$$PR = \frac{\text{Cost per unit of added resource}}{\text{Cost per unit of replaced resource}}$$

Quest. Present value of future investment is calculated by

Ans. **Discounting**

Quest. A line represents the different combinations of two variable inputs used in the production of a given amount of output.

Ans. **Isoquant**

Quest. Isoquant is used in

Ans. **F - F relationship**

Quest. A line indicates all possible combinations of two inputs which can be purchased with a given amount of investment fund

Ans. **Iso-Cost line**

Quest. The line join the end points of Isoquants

Ans. **Ridge line**

Quest. The line by which all the least cost combination points are joined to each other

Ans. **Expansion Path**

Quest. Two or more products when produced in the same production process called

Ans. **Joint product**

Quest. When the increase in one product is directly proportionate to increase the other product is called

Ans. **Complementary Production**

Quest. Crop production and dairy enterprise having

Ans. **Supplementary relationship**

Quest. Fixed Cost + Variable Cost is

Ans. **Total cost**

Quest. The costs, related to fixed resources

Ans. **Fixed Costs**

Quest. Rent, interest on fixed capital, depreciation of building, taxes and wages of the permanent labourers constitute

Ans. **Fixed Costs**

Quest. The cost related to the variable resources and change with the output

Ans. **Variable Costs**

Quest. Gross income - Total Cost is equal to

Ans. **Profit**

Quest. The change in cost associated with an increase of one unit of output.

Ans. **Marginal cost (MC)**

Quest. Variable cost is also known as

Ans. **Prime cost/Input cost**

Quest. Fixed cost is also known as

Ans. **Overhead cost**

Quest. All actual expenses in cash and kind incurred in production by owner operator

Ans. **Cost-A**

Quest. Cost A_1 + rent paid for based in land

Ans. **Cost A_2**

Quest. Cost A_2 + interest on value of owned capital assets (including land)

Ans. **Cost B_1**

Quest. Cost B_1 + rental value of owned land and rent paid by leased in land

Ans. **Cost B_2**

Quest. The total cost of production which includes all cost items, actual as well as imputed.

Ans. **Cost C**

Quest. Cost C is equal to

Ans. **Cost B + imputed value of family human labour**

Quest. Gross Returns – Cost A is

Ans. **Farm business income**

Quest. Gross returns - Cost B is

Ans. **Family labour income**

Quest. Net income is

Ans. **Gross returns - Cost C**

Quest. Benefit Cost Ratio is

Ans. **Gross income/Cost C**

Quest. When farms is classified on the basis of utilization of land and resources, termed as

Ans. **Types of Farming**

Quest. The farming having ≥ 50% income by single enterprise

Ans. **Specialized farming**

Quest. Crop Production + livestock raising is called

Ans. **Mixed farming**

Quest. The farming which has < 50% income by single enterprise

Ans. **Diversified farming**

Quest. Farming in an areas having average annual rainfall of £ 50 cm.

Ans. **Dry farming**

Quest. Natural grazing pattern is known as

Ans. **Ranching**

Quest. When farm is classified on the basis of organizational setup, termed as

Ans. **System of farming**

Quest. The joint agriculture operation by farmer on voluntary basis

Ans. **Cooperative farming**

Quest. The cooperative farming in which Ownership and operations both Individual

Ans. **Cooperative better farming**

Quest. The cooperative farming in which Ownership is individual and operations is collectively

Ans. **Cooperative joint farming**

Quest. The farming in which investment of land and capital is done by big businessperson or capitalist

Ans. **Capitalistic farming**

Quest. Government carries out farming is

Ans. **State farming**

Quest. When farmers follows agricultural practices in their own way and managers and organizers of their farm business

Ans. **Peasant faming**

Quest. A process of deciding in the present what to do in the future about the best combination of crops and live stock to be raised

Ans. **Farm Planning**

Quest. Long-term planning is done for

Ans. **5-10 years**

Quest. A process of estimating costs, returns and net profit of a farm or a particular enterprise

Ans. **Farm Budgeting**

Quest. The basis of farm budgeting is

Ans. **Cost benefit analysis**

Quest. When new variety is recommended, which type of budget should be prepared?

Ans. **Partial budget**

Quest. Net worth is calculated from

Ans. **Balance Sheet**

Quest. The farm accounting/ accountancy is also called as

Ans. **Farm Book keeping**

Chapter 10
Agricultural Statistics

(I) Elements of Statistics

Quest. Mean, Median and Mode are

Ans. **Measures of Central tendency**

Quest. A figure obtained by dividing the sum of all variable by their total number of variables.

Ans. **Averages/Arithmetic Mean**

Quest. Sum of deviation of items from the A.M. is

Ans. **0**

Quest. Which mean is affected by change in origin and scale both?

Ans. **AM**

Quest. Middle most value of the series

Ans. **Median**

Quest. Which one represents median?

Ans. **50th Percentile**

Quest. Most frequently occurred item

Ans. **Mode**

Quest. Relationship between AM, median and Mode in asymmetrical distribution

Ans. **Mode = 3 Median – 2 Mean**

Quest. The best measures of central tendency

Ans. **Arithmetic Mean (AM)**

Quest. The ratio of no. of observations to the sum of the reciprocal of the value of the different observations.

Ans. **Harmonic Mean**

Quest. The order of three averages for a given data

Ans. **AM > GM > HM**

Quest. Mean applied when deals with rate, price and speed of a vehicle

Ans. **HM**

Quest. Mean applied when deals with relative changes eg. Bacterial growth, cell division, population

Ans. **GM**

Quest. The average of the sum of squares of the deviation about mean

Ans. **Variance**

Quest. The degree of scatterness or variation of the variable about a central tendency

Ans. **Disperson**

Quest. MD, SD and Variance are

Ans. **Measures of Disperson/Spread**

Quest. ½ of the interquartile range is

Ans. **Quartile deviation**

Quest. The best measure of Disperson is

Ans. **Standard Deviation (SD)**

Quest. SD is always calculated by

Ans. **AM**

Quest. SD is ranges from

Ans. **0 to ∞**

Quest. The difference between highest and lowest value of the series

Ans. **Range**

Quest. Unit less figure based on two values

Ans. **Range**

Quest. Coefficient of variation calculated by

Ans. **CV = (SD/Mean) x 100**

Quest. The variation used to compare the variability between two series

Ans. **CV**

Quest. Which is not a measure of Disperson?

Ans. **CV**

Quest. The measures of the direction and degree of asymmetry

Ans. **Skewness**

Quest. The formulae of Karl pearson's coefficient of Skewness

Ans. $\mathbf{CSK = \frac{(Mean - Mode)}{\sigma}}$

Quest. Coefficient of skewness for normal distribution is

Ans. **0**

Quest. An idea about the flatness/ peakedness of the curve

Ans. **Kurtosis**

Quest. The term 'Kurtosis' was introduced by

Ans. **Karl Pearson (1906)**

Quest. The curve have $\beta_2 > 3$ or $Y_2 > 0$ is

Ans. **Leptokurtic curve**

Quest. The curve have $b_2 = 3$ or $Y_2 = 0$ is

Ans. **Mesokurtic curve**

Quest. The study the association or degree and deviation between two or more variables.

Ans. **Correlation**

Quest. Correlation lies between

Ans. **-1 to +1**

Quest. Which is used to measure the average relationship between two or more variables?

Ans. **Regression**

Quest. Regression coefficient is independent of

Ans. **Origin**

Quest. The distribution in which Mean > Variance

Ans. **Bionomial distribution**

Quest. The distribution in which Mean = Variance

Ans. **Poison distribution**

Quest. The degree of freedom of Normal distribution

Ans. **n-3**

Quest. The term used to denote chance of happening or not happening of an event.

Ans. **Probability**

Quest. Probability is formulated by

Ans. **Probability =**

$$\frac{\textbf{No. of favourable cases}}{\textbf{Total no. of equally likely cases}}$$

Quest. Probability ranges from

Ans. **0 to 1**

Quest. The test used for comparing two means when sample size is small (up to 30)

Ans. **'T' test**

Quest. Students t test is used when

Ans. **Small samples size and SD is unknown**

Quest. Students t test was proposed by

Ans. **W.S. Gosset**

Quest. To test the proportions and variance, we use

Ans. **'F' test**

Quest. To test the goodness of fit or homogeneity, we use

Ans. **CHI^2 test**

Quest. CHI^2 test was given by

Ans. **Karlpierson**

Quest. When the calculated F is greater than table F value at 5% only, the differences in treatments is considered.

Ans. **Significant**

Quest. With increasing number of error degree of freedom, table F value follow ________ trend.

Ans. **Gradually decreased**

(II) Field Experimentation

Quest. Logical constructions of the experiments in which the degree of uncertainty with which the inference (Result/confusion) on may be well defined.

Ans. **Design of Experiments**

Quest. The objects of comparison, which an experiment has to try in the field for assessing their value.

Ans. **Treatment**

Quest. The 3 basic principles of field experimentation

Ans. **Replication, Randomization and Local control**

Quest. Repeated application of treatments

Ans. **Replication**

Quest. Allocation of treatments to the different experimental units by a random process

Ans. **Randomization**

Quest. Which principle of experimentation eliminates human biases

Ans. **Randomization**

Quest. Local control helps in reducing the

Ans. **Experimental error**

Quest. The transformation required when data not follow normal distribution.

Ans. **Data transformation**

Quest. The most appropriate transformation for percentage

Ans. **Angular transformation**

Quest. Which transformation is applied when mean a variance

Ans. **Square root**

Quest. The hypothesis under test

Ans. **Null hypothesis**

Quest. The variation due to uncontrolled factors

Ans. **Experimental error**

Quest. The error in which hypothesis is true but our test rejects it.

Ans. **Type I error**

Quest. Out of the two types of error in testing, the more severe error is

Ans. **Type II error**

Quest. The simplest experimental design

Ans. **Complete Randomized Design (CRD)**

Quest. The experimental design which provides maximum degree of freedom for error

Ans. **CRD**

Quest. Which design is applied when experimental material are limited and homogenous

Ans. **CRD**

Quest. The error degree of freedom in CRD is formulated as

Ans. **N - t**

Quest. The most commonly used design

Ans. **Randomized Block Design (RBD)**

Quest. RBD is also called as

Ans. **One way elimination of heterogeneity design/Two way classification of ANOVA**

Quest. When fertility gradient in one direction, the statistical design to be used

Ans. **RBD**

Quest. The maximum no. of treatments adopted in RBD

Ans. **20**

Quest. In RBD, the number of blocks is equal to

Ans. **No. of replications (b = r)**

Quest. The error degree of freedom of RBD is formulated as

Ans. **(t-1) (r-1)**

Quest. The design in which fertility gradient is in two way direction

Ans. **Latin Square Design (LSD)**

Quest. LSD is also known as

Ans. **Two way elimination of heterogeneity design/Three way classification of ANOVA**

Quest. In LSD, the no. of row or column or treatment is equal to

Ans. **No. of replications (r = c = t)**

Quest. The optimum number of treatments studied in latin square design

Ans. **5 to 12**

Quest. The error degree of freedom of LSD is formulated as

Ans. **(t-2) (t-1)**

Quest. Which design provides main effects and interactions

Ans. **Factorial RBD**

Quest. The treatment df for 3 factors each at 2 levels is

Ans. **2^3 = 6-1 = 5**

Quest. The technique of reducing the size of replication over a number of blocks at the cost of loosing some informations on same effect

Ans. **Confounding Design**

Quest. Which are unimportant in Confounding Design?

Ans. **Interactions**

Quest. Confounding Design is adopted when the number of treatments is

Ans. **10**

Quest. If an interaction effect is confounded with all the replicates of the treatment

Ans. **Complete/total confounding**

Quest. The most appropriate design, when all factors are not of equally important in experimentation.

Ans. **Split Plot Design (SPD)**

Quest. To study two factors with different level of precision, which design is used

Ans. **Split plot design**

Quest. The factor requires larger units to be applied and may produce larger differences

Ans. **Main plot**

Quest. The error degree of freedom of SPD is formulated as

Ans. **D (r-1) (d-1)**

Quest. In a split plot design, 5 levels of main plot and 4 levels of sub plot treatments studied with 3 replications. What will be the d.f. for error b source?

Ans. **30**

Quest. If sub treatments are laid out in strips then the design is called

Ans. **Strip Plot Design**

Quest. How many no. of error variance are applied in Strip Plot Design

Ans. **3**

Quest. In Strip Plot Design, which one is to be tested with higher precision

Ans. **Interaction**

Chapter 11
Agroforestry

Quest. A system where agriculture and forestry are practised either simultaneously or separately on the same unit of land

Ans. **Agroforestry**

Quest. Agroforestry is a form of

Ans. **Multiple cropping**

Quest. The area under forest land in India

Ans. **67 mha (20.36%)**

Quest. Optimum area under forest required

Ans. **33% of total geographical area**

Quest. Contribution of forest product in world GDP

Ans. **1 %**

Quest. Indian Forest Act was come in existence

Ans. **1927**

Quest. Forest Conservation Act was made in

Ans. **1980**

Quest. Forest school is established at

Ans. **Dehradun**

Quest. The Van Mahotsav Day in India is observed on

Ans. **1 July**

Quest. State having highest forest area in India

Ans. **M.P.**

Quest. Forest type found maximum in India

Ans. **Tropical dry deciduous forest**

Quest. National Research Centre for Agroforestry is situated at

Ans. **Jhansi (1988)**

Quest. International Centre for Research in Agroforestry (ICRAF) is situated at

Ans. **Nairobi, Kenya**

Quest. The most important Agroforestry practice is

Ans. ***Acacia leucophloea* + *Cenchrus setigerus***

Quest. The oldest known agro forestry practice

Ans. **Shifting cultivation**

Quest. Cultivation of Trees + Crops is known as

Ans. **Agri-Silviculture**

Quest. Perennial hedges + crops

Ans. **Alley cropping**

Quest. Fruit trees + crops

Ans. **Agri-horticulture**

Quest. Trees + fruit trees + crops

Ans. **Agri-silvi-horticulture**

Quest. Trees + crops + pasture/animals

Ans. **Agri-silviculture**

Quest. Trees+ pasture/animals

Ans. **Silvi-pasture**

Quest. Fruit trees + honeybees

Ans. **Horti-apiculture**

Quest. Trees + fishes

Ans. **Aqua-forestry**

Quest. Forage trees + pasture

Ans. **Forage forestry**

Quest. Trees + crops during initial years

Ans. **Energy plantation**

Quest. Multiple combination of trees, fruit trees, vegetables atc.

Ans. **Homestead**

Quest. Trees is on boundary + crops

Ans. **Boundary plantation**

Quest. Taungya system means

Ans. **Hill cultivation**

Quest. Most common example of taungya system

Ans. **Planting of Teak in Myanmar**

Quest. Nitrogen fixing tree

Ans. ***Leucaena leucocephala***

Quest. Non leguminous nitrogen fixing trees

Ans. ***Alnus nepalensis***

Quest. Bio-drainage plant

Ans. ***Eucalyptus tereticornis***

Quest. Most suitable woodlot trees in India

Ans. ***Casuarina and Leucaena***

Quest. Fodder producing tree

Ans. ***Prosopis cineraria***

Quest. Fuel wood tree

Ans. ***Albizia lebbeck***

Quest. Green manuring tree

Ans. ***Thespesia populnea***

Quest. Shifting cultivation causes

Ans. **Deforestation**

Quest. Miracle forest tree (as it provides fodder, fuel, pulpwood and timber)

Ans. **Subabul**

Quest. Fast growing forest tree species

Ans. ***Eucalyptus sp.***

Quest. Multipurpose tree species

Ans. ***Albizia lebbeck***

Quest. The most appropriate and effective type of crop cultivation in forests

Ans. **Intercropping**

Quest. Ratanjot and Karanj are

Ans. **Biofuel plants**

Quest. Oil percentage in Ratanjot (*Jatropha sp.*)

Ans. **35% (from seed)**

Quest. Spacing maintained between hedge row intercropping in alley cropping

Ans. **4-8 meter**

Quest. Tree species suitable for alley cropping

Ans. ***Cassia siamea, Leucaena* and *Sesbania***

Quest. Forestry outside the conventional forests which primarily aim at providing continuous flow of goods and services for the benefit of people

Ans. **Social forestry**

Quest. A forest system which promote commercial tree growing by farmers on their own land

Ans. **Farm Forestry**

Quest. Pollarding is done at

Ans. **2 m height from ground**

Quest. A belt of trees and or shrubs maintained for the purpose of shelter from wind, sun, snow drift, etc.

Ans. **Shelterbelts**

Quest. A protective plantation in a certain area, against strong winds. It is usually comprised of a few rows of trees (or shrubs)

Ans. **Wind breaks**

Quest. Raising of forests of public or community land

Ans. **Community forestry**

Quest. The ratio of height, width and length in shelterbelt system

Ans. **1:25:10 meter**

Quest. A process in which the branch of a plant is cut off in order to produce a flush of new shoots

Ans. **Pollarding**

Quest. The main stem of a tree is called

Ans. **Bol**

Quest. Full form of ICARF

Ans. **International Centre for Research in Agroforestry**

Chapter 12

Environmental Science and Ecology

Quest. The sum total condition in which organisms live is called as

Ans. **Environment**

Quest. A self supporting community - plants and animals interacting with each other and the non-living environment to provide a balanced system is a

Ans. **Ecosystem**

Quest. The word 'ecosystem' was coined by

Ans. **A.G.Tansley**

Quest. The region existing between two ecosystems which contain species of both ecosystems is

Ans. **Ecotone**

Quest. The study of interactions between living organism and environment is called as

Ans. **Ecology**

Quest. The term ecology was introduced by

Ans. **Hackel**

Quest. Large portions of the earth with similar climate, soil, plant and animal life community is known as

Ans. **Biosphere**

Quest. The earth contains a thin region known as the biosphere, in which life exists. The three parts of the biosphere are

Ans. **Atmosphere, hydrosphere and lithosphere**

Quest. Sphere of Water or 70% of global is occupied by

Ans. **Hydrosphere**

Quest. The characteristics of the type of environment where an organism normally lives?

Ans. **Habitat**

Quest. All the populations of the different species living and interacting in the same ecosystem?

Ans. **Community**

Quest. The variety of living organisms (flora and fauna) is called as

Ans. **Biodiversity**

Quest. The environment which includes producers, consumers and decomposers?

Ans. **Biotic environment**

Quest. Abiotic environment does not include

Ans. **Plants**

Quest. The environment which has been modified by human activities is called

Ans. **Anthropogenic environment**

Quest. The group of organisms which convert light into food are called

Ans. **Autotrophs**

Quest. The plants, which produce food themselves through photosynthesis?

Ans. **Phototrophs**

Quest. The ecosystem component, feed on producer or consumers?

Ans. **Heterotrophs**

Quest. The Heterotrophs, feed on plants and called primary consumers?

Ans. **Herbivores**

Quest. The Heterotrophs, feed on meats and called secondary consumers?

Ans. **Carnivores**

Quest. The main constituent of CNG is

Ans. **Methane**

Quest. The Heterotrophs, feed on both plants and meats?

Ans. **Omnivores**

Quest. The base of the food chain in the ocean is the

Ans. **Phytoplankton**

Quest. Primary consumers in aquatic system are?

Ans. **Zooplanktons**

Quest. The ecosystem component, which break down dead organic matter and wastes?

Ans. **Decomposers**

Quest. Decomposers include

Ans. **Bacteria and Fungi**

Quest. In which ecosystem, producers are of large size

Ans. **Grassland ecosystem**

Quest. Natural resources which are renewable

Ans. **Water and wood**

Quest. Non Renewable resources are

Ans. **Minerals, fossil, fuels.**

Quest. The main constituent of LPG is

Ans. **Butane**

Quest. The source of energy for all plants is

Ans. **Sun**

Quest. The ecological factors, related to soil and substratum, are called

Ans. **Edaphic Factor**

Quest. The term used to refer the weight of all the organisms at a tropical level.

Ans. **Biomass**

Quest. The unit of energy is

Ans. **Joule**

Quest. Which is not included under biomass?

Ans. **Water**

Quest. Plants which grow on other plants are called

Ans. **Epiphytes**

Quest. Increasing industrialisation is causing much danger to man's life by

Ans. **Polluting the environment**

Quest. Major pollutants that contribute to 90% of global air pollution

Ans. **CO and CO_2**

Quest. Global warming focuses on an increase in the level of which gas in the atmosphere?

Ans. **Carbon dioxide**

Quest. Colourless and odourless air pollutant is

Ans. **SO_2**

Quest. Most poisonous pollutant in water

Ans. **Arsenic**

Quest. Most commonly used disinfectant in water purification

Ans. **Chlorine**

Quest. The test which has self purification capacity of water body

Ans. **BOD (Biochemical Oxygen Demand) test**

Quest. Materials that cause BOD include

Ans. **Wood, animal wastes, sewage**

Quest. Permissible limit of iron in drinking water

Ans. **1 ppm**

Quest. A disease caused by mercury (Hg) poisoning of water at Japan in 1953

Ans. **Minamata**

Quest. *Manimata* disease is due to

Ans. **Mercury toxicity**

Quest. *Itai-Itai* disease is due to

Ans. **Cadmium (Cd) toxicity**

Quest. The chief green house gases are

Ans. **CO_2 and CH_4**

Chapter 13
About ICAR and IARI

I.C.A.R.

- The highest body controlling agricultural research and education in India is" **Indian Council of Agricultural Research** (ICAR)."
- It was established on July 16, 1929 with the name "Imperial Council of Agricultural Research" under the Societies Registration Act, 1860 in pursuance of the report of the Royal Commission on Agriculture.
- ICAR headquarters at Krishi Bhavan, New Delhi.
- The ICAR was bestowed with the **King Baudouin Award** in 1989 for its valuable contribution in the Green Revolution. Again awarded King Baudouin Award in 2004 for research and development efforts made under partnership in Rice Wheat Consortium.
- First Director-General was **Dr. B.P. Pal** (1965)
- Union Minister of Agriculture is the ex-officio President of the ICAR Society. (Present- **Sharad Pawar**)
- New Director-General of ICAR: **Dr. S. Ayyappan** (from 01.01.2011)
- 4 Deemed universities are part of the ICAR.

I.A.R.I.

- 1905: Agricultural Research Institute was established at Pusa, Bihar by Lord Curzon. The land was donated by Mr. Phipps of USA after whom the place was named as Pusa. The Phipps laboratory in division of Soil Science and Agricultural Chemistry.
- 1911: Renamed as Imperial Agricultural Research Institute.
- 1923: Institute started offering Diploma of Associateship.
- 1934: Major Earth quake damages the buildings at Pusa.

- 1936: Shifted to New Delhi.
- 1936: **B. Vishwanath** became the first Indian Director of the Institute.
- 1946: The Diploma of Associateship was recognised equivalent to M.Sc.
- 1947: Name has been changed from Imperial Agricultural Research Institute to **Indian Agricultural Research Institute**.
- 1958: Recognized as "Deemed University" under UGC Act of 1956.

Chapter 14

Agricultural Research, Education and Extension

DEEMED UNIVERSITIES - 4

1.	Indian Agricultural Research Institute (IARI)	:	New Delhi
2.	National Dairy Research Institute (NDRI)	:	Karnal
3.	Indian Veterinary Research Institute (IVRI)	:	Izatnagar
4.	Central Institute on Fisheries Education (CIFE)	:	Mumbai

NATIONAL RESEARCH INSTITUTES - 45

1.	Central Rice Research Institute (CRRI)	:	Cuttack
2.	Vivekananda Parvatiya Krishi Anusandhan Sansthan (VPKAS)	:	Almora
3.	Indian Institute of Pulses Research (IIPR)	:	Kanpur
4.	Central Tobacco Research Institute (CTRI)	:	Rajahmundry
5.	Indian Institute of Sugarcane Research (IISR)	:	Lucknow
6.	Sugarcane Breeding Institute (SBI)	:	Coimbatore
7.	Central Institute of Cotton Research (CICR)	:	Nagpur
8.	Central Research Institute for Jute and Allied Fibres (CRIJAF)	:	Barrackpore
9.	Indian Grassland and Fodder Research Institute (IGFRI)	:	Jhansi
10.	Indian Institute of Horticultural Research (IIHR)	:	Bangalore
11.	Central Institute of Sub Tropical Horticulture (CISTH)	:	Lucknow

12.	Central Institute of Temperate Horticulture (CITH)	:	Srinagar
13.	Central Institute of Arid Horticulture (CIAR)	:	Bikaner
14.	Indian Institute of Vegetable Research (IIVR)	:	Varanasi
15.	Central Potato Research Institute (IPRI)	:	Shimla
16.	Central Tuber Crops Research Institute (CTCRI)	:	Trivandrum
17.	Central Plantation Crops Research Institute (CPCRI)	:	Kasargod
18.	Central Agricultural Research Institute (CARI)	:	Port Blair
19.	Indian Institute of Spices Research (IISR)	:	Calicut
20.	Central Soil and Water Conservation Research & Training Institute (CSWCRTI)	:	Dehradun
21.	Indian Institute of Soil Sciences (IISS)	:	Bhopal
22.	Central Soil Salinity Research Institute (CSSRI)	:	Karnal
23.	ICAR Research Complex for Eastern Region including Centre of Makhana	:	Patna
24.	Central Research Institute of Dryland Agriculture (CRIDA)	:	Hyderabad
25.	Central Arid Zone Research Institute (CAZRI)	:	Jodhpur
26.	ICAR Research Complex	:	Goa
27.	ICAR Research Complex for NEH Region	:	Barapani
28.	National Institute of Abiotic Stress Management (NIASM)	:	Malegaon
29.	Central Institute of Agricultural Engineering (CIAE)	:	Bhopal
30.	Central Institute on Post-harvest Engineering and Technology (CIPET)	:	Ludhiana
31.	Indian Institute of Natural Resins and Gums (IINRG)	:	Ranchi
32.	Central Institute of Research on Cotton Technology (CIRCT)	:	Mumbai
33.	National Institute of Research on Jute & Allied Fibre Technology (NIRJAFT)	:	Kolkata
34.	Indian Agricultural Statistical Research Institute (IASRI)	:	New Delhi
35.	Central Sheep and Wool Research Institute (CSWRI)	:	Avikanagar
36.	Central Institute for Research on Goats (CIRG)	:	Makhdoom
37.	Central Institute for Research on Buffaloes (CIRB)	:	Hissar
38.	National Institute of Animal Nutrition and Physiology (NIANP)	:	Bangalore
39.	Central Avian Research Institute (CARI)	:	Izatnagar
40.	Central Marine Fisheries Research Institute (CMFRI)	:	Kochi

41.	Central Institute Brackishwater Aquaculture (CIBA)	:	Chennai
42.	Central Inland Fisheries Research Institute (CIFRI)	:	Barrackpore
43.	Central Institute of Fisheries Technology (CIFT)	:	Cochin
44.	Central Institute of Freshwater Aquaculture (CIFA)	:	Bhubneshwar
45.	National Academy of Agricultural Research & Management (CAARM)	:	Hyderabad

NATIONAL RESEARCH CENTRES - 17

1.	National Research Centre on Plant Biotechnology (NRCPB)	:	New Delhi
2.	National Centre for Integrated Pest Management (NCIPM)	:	New Delhi
3.	National Research Centre for Litchi (NRCL)	:	Muzaffarpur
4.	National Research Centre for Citrus (NRCC)	:	Nagpur
5.	National Research Centre for Grapes (NRCG)	:	Pune
6.	National Research Centre for Banana (NRCB)	:	Trichi
7.	National Research Centre Seed Spices (NRCSS)	:	Ajmer
8.	National Research Centre for Pomegranate (NRCP)	:	Solapur
9.	National Research Centre on Orchids (NRCO)	:	Pakyong, Sikkim
10.	National Research Centre Agroforestry (NRCA)	:	Jhansi
11.	National Research Centre on Camel (NRCC)	:	Bikaner
12.	National Research Centre on Equines (NRCE)	:	Hisar
13.	National Research Centre on Meat (NRCM)	:	Hyderabad
14.	National Research Centre on Pig (NRCP)	:	Guwahati
15.	National Research Centre on Yak (NRCY)	:	West Kemang
16.	National Research Centre on Mithun (NRCM)	:	Medziphema
17.	National Centre for Agril. Economics & Policy Research (NCAEPR)	:	New Delhi

NATIONAL BUREAUX - 6

1.	National Bureau of Plant Genetics Resources (NBPGR)	:	New Delhi
2.	National Bureau of Agriculturally Important Micro-organisms (NBAIM)	:	Mau
3.	National Bureau of Agriculturally Important Insects (NBAII)	:	Bangalore

4.	National Bureau of Soil Survey and Land Use Planning (NBSSLUP)	:	Nagpur
5.	National Bureau of Animal Genetic Resources (NBAGR)	:	Karnal
6.	National Bureau of Fish Genetic Resources (NBFGR)	:	Lucknow

INTERNATIONAL ORGANIZATIONS OF CROP IMPROVEMENT

CIAT	International Centre for Tropical Agriculture	:	Cali, Columbia
CIFOR	Center for International Forestry Research	:	Jakarta, Indonesia
CIMMYT	International Centre for Wheat and Maize Improvement	:	Baton, Mexico
CIP	International Potato Centre	:	Lima, Peru
IBPGR	International Board for Plant Genetic Resources	:	Rome, Italy
ICARDA	International Centre for Agricultural Research in the Dry Areas	:	Alleppo, Syria
ICGES	International Centre for Genetic Engineering and Biotechnology	:	Triesta, Italy
ICRAF	International Centre for Research in Agroforestry	:	Nairobi, Kenya
ICRISAT	International Crops Research Institute for the Semi-Arid Tropics	:	Hyderabad, India
IFPRI	International Food Policy Research Institute	:	Washington, USA
IITA	International Institute of Tropical Agriculture	:	Ibadan, Nigeria
IIMI	International Irrigation Management Institute	:	Colombo, Srilanka
ILRI	International Livestock Research Institute	:	Nairobi, Kenya
INSFFER	International Network on Soil Fertility and Fertilizer Evaluation on Rice	:	New Delhi, India
IPGRI	International Plant Genetic Resource Institute	:	Rome, Italy
ISNAR	International Service for National Agricultural Research	:	Netherlands
IRRI	International Rice Research Institute	:	Manila, Phillipines
IWMI	International Water Management Institute	:	Columbo, Sri Lanka
WFC	World Fish Centre	:	Bayan Lepas, Malaysia
WARDA	West African Rice Development Association	:	Monrovia, Liberia

STATE AGRICULTURE UNIVERSITIES - 50

1.	Acharya NG Ranga Agricultural University	:	Rajendra Nagar, Hyderabad (AP)
2.	Anand Agricultural University	:	Anand, Gujarat
3.	Assam Agricultural University	:	Jorhat, Assam
4.	Bidhan Chandra Krishi Viswavidyalaya	:	Mohanpur, Nadia, (WB)
5.	Birsa Agricultural University	:	Ranchi, Jharkhand
6.	Central Agricultural University	:	Imphal, Manipur
7.	Chandra Shekar Azad University of Agriculture & Technology	:	Kanpur (UP)
8.	Chaudhary Charan Singh Haryana Agricultural University	:	Hisar, Haryana
9.	CSK Himachal Pradesh Krishi Vishvavidyalaya	:	Palampur, Himachal Pradesh
10.	Dr Balasaheb Sawant Konkan Krishi Vidyapeeth	:	Dapoli Distt, Maharashtra
11.	Dr Panjabrao Deshmukh Krishi Vidyapeeth	:	Akola, Maharashtra
12.	Dr Yashwant Singh Parmar Univ of Horticulture & Forestry	:	Solan, Himachal Pradesh
13.	Govind Ballabh Pant University of Agriculture & Technology	:	Pantnagar, Uttaranchal
14.	Guru Angad Dev Veterinary and Animal Science University	:	Ludhiana, Punjab
15.	Indira Gandhi Krishi Vishwavidyalaya	:	Raipur, Chhattisgarh
16.	Jawaharlal Nehru Krishi Viswavidyalaya	:	Jabalpur (MP)
17.	Junagadh Agricultural University	:	Junagad, Gujarat
18.	Kerala Agricultural University	:	Trichur, Kerala
19.	Maharana Pratap Univ. of Agriculture & Technology	:	Udaipur, Rajasthan
20.	Maharashtra Animal Science & Fishery University	:	Nagpur, Maharashtra
21.	Mahatma Phule Krishi Vidyapeeth	:	Rahuri, Maharashtra
22.	Marathwada Agricultural University	:	Parbhani, Maharashtra
23.	Narendra Deva University of Agriculture & Technology	:	Faizabad (UP)

24.	Navsari Agricultural University	:	Navsari, Gujarat
25.	Orissa Univ. of Agriculture & Technology	:	Bhubaneshwar, Orissa
26.	Punjab Agricultural University	:	Ludhiana, Punjab
27.	Rajasthan Agricultural University	:	Bikaner,Rajasthan
28.	Rajendra Agricultural University	:	Pusa, Bihar
29.	Sardarkrushinagar-Dantiwada Agricultural University	:	Sardar Krushinagar, Gujarat
30.	Sardar Ballabh Bhai Patel Univ. of Agriculture & Technology	:	Modipuram (UP)
31.	Sher-E-Kashmir Univ of Agricultural Sciences & Technology	:	Railway Road, Jammu
32.	Sher-E-Kashmir Univ of Agricultural Sciences & Technology of Kashmir	:	Srinagar, J &K
33.	Sri Venkateswara Veterinary University	:	Tirupati
34.	Tamil Nadu Agricultural University	:	Coimbatore, Tamil Nadu
35.	Tamil Nadu Veterinary & Animal Science University	:	Chennai, Tamil Nadu
36.	University of Agricultural Sciences	:	Bangalore, Karnataka
37.	University of Agricultural Sciences	:	Dharwad, Karnataka
38.	U.P. Pandit Deen Dayal Upadhaya Pashu Chikitsa Vigyan Vishwa Vidhyalaya evam Go Anusandhan Sansthan	:	Mathura (UP)
39.	Uttar Banga Krishi Viswavidyalaya	:	Coach Bihar (WB)
40.	West Bengal University of Animal & Fishery Sciences	:	Kolkata, (WB)
41.	Karnataka Veterinary, Animal and Fisheries Sciences University	:	BIDAR, Karnataka
42.	University of Agricultural Sciences	:	Raichur, Karnataka
43.	University of Horticultural Sciences	:	Bagalkot, Karnataka
44.	Andhra Pradesh Horticultural University	:	Tadepalligudem, Andhra Pradesh
45.	Rajmata Vijay Raje Sciendia Krishi Vishwa-vidyalaya	:	Gwalior (MP)

New Approved Universities

46.	Bihar Agricultural University	:	Bhagalpur, Bihar
47.	Kerala University of Fisheries and Oceanography	:	Kochi (Kerala)
48.	Manyavar Shri Kanshi Ramji Agriculture Technology	:	Banda (U.P.)
49.	Rajasthan University of Veterinary Sciences	:	Bikaner (Rajasthan)
50.	Tamilnadu Horticultural University	:	Krishnagiri (TN)

Chapter 15
Useful Information

FATHERS OF DIFFERENT DISCIPLINES

	Father of		*Name*
☆	Agronomy	:	Pietro Decrescenzi
☆	Agro meteorology	:	D. N. Walia
☆	Agricultural chemistry	:	Justus von Liebig
☆	Bacteriology	:	Leuwenhoek
☆	Biochemistry	:	Justus von Liebig
☆	Cooperative movement in India	:	F. Nicholson
☆	Extension education	:	A. Seaman/Leagnes
☆	Experimental genetics	:	Thomas Hunt Morgan
☆	Field plot experiment	:	J. B. Boussingault
☆	Fruit and vegetable preservation	:	M. Nicholas Apart
☆	Genetics	:	Gregor Johann Mendel
☆	Green revolution	:	Dr. N. E. Borlaug
☆	Green revolution in India	:	M.S.Swaminathan
☆	Golden revolution in India	:	Dr. K.C. Chadha
☆	Golden rice	:	Dr. Ingo Potrykus
☆	Hybrid rice	:	Yuan Long Ping
☆	Hybrid cotton	:	C.T. Patel
☆	Indian plant pathology	:	E.J. Butler
☆	Indian Rust	:	Dr. K.C. Mehta
☆	Microbiology	:	Louis Pasture
☆	Modern Genetics	:	T.H. Morgan

	Father of		*Name*
☆	Mutation Theory	:	Hugo de vries
☆	Ornamental Gardening	:	M. S. Randhawa
☆	Plant Pathology	:	Anton De Bary
☆	Plant Physiology	:	Stephen Hales
☆	Pedology	:	V.V. Dokuchalev
☆	Plant Tissue Culture	:	G. Haberlandt
☆	Sociology	:	Auguste compte
☆	Statistics	:	R.A. Fisher
☆	Soil Science	:	Dokuchalev
☆	Soil Microbiology	:	S.N. Winogradsky
☆	Super Rice	:	Dr. G.H. Khush
☆	Tillage and Weeds	:	Jethro Tull
☆	White Revolution	:	Dr. Varghese Kurien

BOTANICAL NAME OF CROPS

	Crops		*Botanical Name*
☆	**Cereal Crops**		
	Rice	:	*Oryza sativa* L.
	Wheat	:	*Triticum aestivum* L.
	Maize	:	*Zea mays* L.
	Bajra/Pearlmillet	:	*Pennisetum typhoides/P. glaucum* L.
	Sorghum/Jowar	:	*Sorghum bicolor/S. vulgare* L. Moench
	Barley	:	*Hordeum vulgare* L.
	Triticale	:	*Secale cereal*
	Buckwheat/Pseudo cereal	:	*Fagopyrum esculentum*
☆	**Millet Crops**	:	
	Cheena/Proso millet	:	*Panicum miliacearum*
	Foxtail/Italian/Jerman millet/ Kakun	:	*Seteria italica* L. Beauv.
	Kodo/Coarsest millet	:	*Paspulum scrobiculatum* L.
	Little millet	:	*Panicum sumatrense*

Crops		Botanical Name
Madua/Ragi/Finger millet	:	*Eleusine coracana* Gaertn
Sawan/Barnyard millet	:	*Echinochloa frumentance* L.
☆ **Pulse Crops**	:	
Gram/Chickpea/Bengal gram	:	*Cicer aeritinum* L.
Field Pea/Grain pea	:	*Pisum sativum var. arvense*
Arhar/Pigeon pea/Red gram	:	*Cajanus cajan* L. Millsp.
Soybean	:	*Glycine max* L. Merril
Black gram/Urdbean	:	*Vigna mungo/Phaseolus mungo* L. Hepper
Green gram/Moong/Moongbean	:	*Vigna radiate/Phaseolus aureus* L. Wilczek
French bean/Rajmash	:	*Phaseolus vulgaris*
Indian Cowpea/Lobia	:	*Vigna unguiculata/V. sinensis* L.
Lentil	:	*Lens esculantum/L. culinaris* Moench
Lathyrus/Chickling pea/Grasspea	:	*Lathyrus sativus*
Mothbean	:	*Vigna/Phaseolus aconotifolia*
Horse gram/Kulthi	:	*Macrotyloma uniflorum*
☆ **Edible Oilseed Crops**	:	
Groundnut/Peanut/Monkeynut	:	*Arachis hypogea* L.
Sunflower	:	*Helianthus annus* L.
Safflower	:	*Carthamus tinctorius* L.
Rapseed and Mustard	:	*Brassica spp.* L.
Sesamum/Til	:	*Sesamum indicum* L.
Niger	:	*Guzotta abssicinia*
☆ **Non edible Oilseed Crops**	:	
Castor	:	*Ricinus communis* L.
Linseed/Flex	:	*Linnum ussitatisimum* L.
☆ **Fiber Crops**	:	
Cotton	:	*Gossipium spp.*
Jute/tita pat	:	*Corchorus capsularis*
Sunhemp	:	*Crotolaria juncea* L.
☆ **Forage Crops**	:	
Berseem	:	*Trifolium alexandrinum* L.
Lucerne/Alfalfa	:	*Medicago sativa* L.

	Crops		Botanical Name
	Oat	:	*Avena sativa* L.
	Napier grass	:	*Pennisetum purpureum* L.
	Clusterbean/Gaur	:	*Cymopsis tetragonalaba* L.
☆	**Sugar Crops**	:	
	Sugarcane/Cane	:	*Saccharum officinarum* L.
	Sugarbeet	:	*Beta vulgaris* L.
☆	**Tuber Crops**	:	
	Potato	:	*Solanum tuberosum* L.
	Tapioca	:	*Manihot utilissima*
☆	**Stimulate Crops**	:	
	Tobacco	:	*Nicotiana spp.*
	Opium	:	*Papaver somniferum*
☆	**Medicinal Crops**	:	
	Safed musli	:	*Chlorophytum borivilianum*
	Ashwagandha/Winter cherry	:	*Withania somnifera*
	Rouvolfia/Sarpagandha	:	*Rouvolfia serpentina*
	Isabgol	:	*Plantago ovata*
	Butch	:	*Acorus calamus*
	Bramhi	:	*Bacopa morriei*
	Nux vomica	:	*Strychnos Nuxvomica*
☆	**Aromatic Crops**	:	
	Lemon grass	:	*Cymbopogan flexuasus*
	Mentha/Mint	:	*Menthe arvensis*
	Khus/Vetivar	:	*Vetiveria zizanoides*
	Citronella	:	*Cymbopogan winterianus*
	Tulsi/Basil	:	*Ocimum sanctum*
☆	**Fruit Crops**	:	
	Kiwi fruits	:	*Actinidia chinensis*
	Bael	:	*Aegle marmelos*
	Custard apple	:	*Annona squamosa*
	Pineapple	:	*Annanas comosus*
	Jackfruit	:	*Autocarpus heterophyllus*
	(Kair)	:	*Capparis decidue*

Crops		*Botanical Name*
Papaya	:	*Carica papaya*
Karonda	:	*Carissa carandus*
Pecanut	:	*Carya illinoensis*
Lime	:	*Citrus aurantifolium*
Kinnow	:	*Citrus deliciosa*
Lemon	:	*Citrus limon*
Orange	:	*Citrus reticulata*
Sweet orange	:	*Citrus sinensis*
Aonla	:	*Emblica officinalis*
Wood apple	:	*Feronia limonia*
Fig	:	*Ficus carica*
Strawberry	:	*Fragaria sp.*
Phalsa	:	*Grewia subinaequalis*
Walnut	:	*Juglans regia*
Litchi	:	*Litchi chinensis*
Apple	:	*Malus domestica*
Mango	:	*Mangifera indica*
Mulberry	:	*Morus sp.*
Banana	:	*Musa paradisiaca*
Date palm	:	*Phoenix dactylifera*
Almond	:	*Prunus amygdalus*
Apricot	:	*Prunus armeniaca*
Pear	:	*Prunus communis*
Plum	:	*Prunus domestica*
Peach	:	*Prunus persica*
Guava	:	*Psidium guajava*
Pomegranate	:	*Punica granatum*
Raspberry	:	*Rubus idaeus*
Jamun	:	*Syzygium cumini*
Tamarind	:	*Tamarindus indica*
Ber	:	*Zizyphus mauritiana*
Grape	:	*Vitis vinifera*

	Crops		*Botanical Name*
☆	**Vegetable Crops**	:	
	Onion	:	*Allium cepa*
	Garlic	:	*Allium sativum*
	Elephant foot yam	:	*Amorphophyllus campanulatus*
	Asparagus	:	*A. officinalis*
	Beet root	:	*Beeta vulgaris*
	Palak	:	*B. vulgaris var. bengalensis*
	Spinach	:	*Spinacea oleraceae*
	Sweet Potato	:	*Ipomea batatas*
	Cabbage	:	*Brassica oleracea var. capitata*
	Cauliflower	:	*B. o. var. botrytis*
	Brussel's Broccoli	:	*B. o. var. gemmifera*
	Knol-khol	:	*B. caulorapa*
	Turnip	:	*B. rapa*
	Raddish	:	*Raphanus sativus*
	Cucumber	:	*Cucumis sativus*
	Musk melon	:	*Cucumis melo*
	Snap melon (foot)	:	*Cucumis melo var. momordica*
	Long melon (Kakri)	:	*C. melo var. utillisium*
	Gherkin	:	*C. anguria*
	Water melon	:	*Citrullus latanus*
	Round melon	:	*C. l. var. fistulosus*
	Pumpkin	:	*Cucurbita moschata*
	Bottle gourd	:	*Lagenaria siceraria*
	Ridge gourd	:	*Luffa acutangula*
	Sponge gourd	:	*L. cylindrica*
	Pointed gourd	:	*Trichosanthus dioca*
	Snake gourd	:	*T. anguina*
	Ash gourd	:	*Benincasa hispida*
	Ivy gourd	:	*Coccinia indica*
	Spine gourd	:	*Momordica chinensis*
	Bitter gourd	:	*M. charantia*

	Crops		*Botanical Name*
	Peas	:	*Pisum sativum var hartense*
	French bean	:	*Phaseolus vulgaris*
	Cluster bean	:	*Cymopsis tetragonalabus*
	Cowpea	:	*Vigna unguiculata*
	Fenugreek	:	*Trigonella foenugraecum*
	Okra	:	*Abelmoschus esculantus*
	Potato	:	*Solanum tuberosum*
	Tomato	:	*S. lycopersicon*
	Brinjal	:	*S. melongena*
	Chilli	:	*Capsicum annum*
	Sweet pepper	:	*C. annum*
	Carrot	:	*Daucus carota*
	Coriander	:	*Coriandrum sativum*
	Celery	:	*Apium graveolens*
☆	**Flower Crops**	:	
	Rosa	:	*Rosa indica*
	Chrysanthemum	:	*Chrysanthemum spp*
	Gladiolus	:	*Gladiolus spp.*
	Carnation	:	*Dianthus spp.*
	Marigold	:	*Tagetes spp*
	Tuberose	:	*Polianthes tuberose*
	Dahlia	:	*Dahlia pinnata*
	Jasmine	:	*Jasminum spp.*
	Bougainvillea	:	*Bougainvillea spp.*

FAMOUS NAME OF CROPS

	Famous Name		*Crops*
☆	King of cereals	:	Wheat
☆	Queen of cereals	:	Maize
☆	King of pulses	:	Chickpea
☆	Queen of pulses	:	Pea
☆	King of oilseeds	:	Groundnut

	Famous Name		*Crops*
☆	Queen of oilseeds	:	Sesame (Til)
☆	Coarsest of all food grains	:	Kodo (*Paspulum scrobiculatum)*
☆	King of fruits	:	Mango
☆	Queen of fruits	:	Pineapple
☆	King of temperate fruits	:	Apple
☆	King of spices	:	Black Pepper
☆	Queen of spices	:	Cardamom
☆	King of vegetables	:	Potato
☆	Queen of vegetables	:	Okra
☆	Wonder crop	:	Soybean
☆	King of fodder crops	:	Berseem
☆	Queen of fodder crops	:	Lucerne
☆	King of Arid and semi fruits	:	Ber
☆	National fruits of India	:	Mango
☆	Wonder tree	:	Neem
☆	Bio energy plant	:	Jatropha
☆	King of flower crops	:	Rose
☆	Queen of flower crops	:	Gladiolus
☆	Adams fig	:	Banana
☆	Oldest cultivated tropical fruits	:	Banana
☆	Tree of heaven	:	Coconut
☆	King of nut crops	:	Walnut
☆	Queen of nut crops	:	Peanut
☆	White gold of America	:	Cotton
☆	Yellow jewel of America	:	Soybean
☆	Backbone of America	:	Maize
☆	Sugar bowl	:	Cuba

TERMS AND ASSOCIATED CROPS

	Terms		*Associated Crops*
☆	Curing	:	Tobacco, Tea
☆	Nipping	:	Gram

	Terms		*Associated Crops*
☆	Wrapping	:	Sugarcane
☆	Dapog seedling	:	Rice seedling
☆	De- suckering	:	Tobacco
☆	De- tasseling	:	Maize
☆	Pegging	:	Groundnut
☆	Retting	:	Jute
☆	Ginning	:	Cotton
☆	Topping	:	Cotton
☆	Arrowing	:	Sugarcane
☆	Ratooning	:	Sugarcane
☆	Parboiling	:	Rice
☆	Earthing-up	:	Potato, Sugarcane

SEED RATE and NATIVITY OF CROPS

(I) Recommended Seed Rate of Field Crops

	Crops		*Seed Rate (kg/ha)*
☆	Rice		
	a) Transplanting	:	50 - 60
	b) Broadcasting	:	80 - 100
☆	Wheat	:	100 - 125
☆	Maize		
	a) Hybrid	:	20 - 25
	b) Composite	:	15 - 20
☆	Sorghum, Moong, Arhar	:	12 - 15
☆	Pearlmillet	:	2 - 3
☆	Gram	:	60 - 80
☆	Field Pea	:	75 - 100
☆	Urd, Cowpea, Sunhemp	:	20 - 25
☆	Lathyrus		
	a) Pure crop	:	40 - 50
	b) Mixed crop	:	8 - 10

	Crops		Seed Rate (kg/ha)
☆	Lentil, Linseed	:	30 - 40
☆	Soybean	:	70 - 80
☆	Safflower	:	15
☆	Sunflower, Sugarbeet, Jute	:	8 - 10
☆	Groundnut		
	a) Bunch type	:	100 - 120
	b) Spreading type	:	80 - 100
☆	Til	:	3 - 4
☆	Rapseed and Mustard		
	a) Pure cropping	:	4 - 6
	b) Mixed cropping	:	2 - 3
☆	Castor	:	10
☆	Cotton	:	10 - 12
☆	Hybrid cotton, Tobacco	:	2.5 - 3
☆	Potato	:	10 - 15 qt
☆	Kodo	:	6 - 8
☆	Lucerne	:	20
☆	Berseem		
	a) Diploid spp	:	20 - 25
	b) Tetraploid spp	:	30 - 35
☆	Fodder maize	:	40 - 60
☆	Fodder Bajra	:	20 - 30
☆	Oat		
	a) Small seeded	:	80 - 100
	b) Bold seeded	:	100 - 120

(II) Nativity of Field Crops

	Crops		Nativity
☆	Rice, Sugarcane	:	South East Asia
☆	Wheat, Barley, Buckwheat, Gram, Lucerne	:	South West Asia
☆	Soybean, Rapseed and Mustard, Tea	:	China

	Crops		*Nativity*
☆	Tobacco	:	America
☆	Maize, Teosinate	:	Mexico
☆	Potato, Tomato	:	Peru
☆	Linseed	:	Afghanistan
☆	Sunflower,	:	USA
☆	Arhar, Mung, Urd, Cotton, Jute, Kodo, Kutki, Oat, Mango	:	India
☆	Sorghum, Bajra, Sunhemp, Sesamum, Cowpea, Castor, Clusterbean	:	Africa
☆	Groundnut	:	Brazil
☆	Berseem	:	Egypt
☆	Napier grass	:	Rhodesia

IMPORTANT VARIETIES OF FIELD CROPS

☆	Paddy	:	Aaditya, Purnima, IR36, 64, 20, MTU1001, 1010, Indira sona, Kranti, Mahamaya, Safri17, Bumleshwari, Pusa basmati-1, Shyamla, etc.
☆	Wheat	:	Lok-1, C306, HW 2004, WH 147, Sujata, GW 173, 273, Kanchan, Raj.
☆	Maize	:	Ganga-1, 3, 5, 101, Ganga safed-2, Ranjit, Himalaya, VL-54, Ganga-4., Navjot, Chandan makka-3, Chandan safed makka-1 etc.
☆	Kodo	:	GPUK-3, ICCK 737, IPS 147-1, JK 1, 155, Pali.
☆	Kutki	:	PRC3, IGL4, 10
☆	Kulthi	:	BK-1, AK-21, JND-2
☆	Ragi	:	VL-147, PR 202, HR-374
☆	Arhar	:	Type-21, Prabhat, UPAS-120, pragati, Asha, Gwalior-3, Bahar, Rajiv Lochan
☆	Moong	:	Pairy moong 2, pragya, Pusa Baisakhi, JM-721, K-851, PDM-1, 3,11
☆	Urd	:	Pant U-30, JU-2, Type-9, Gwalior-2, Sarla, Barkha, Prabha, CO-1
☆	Gram	:	JG-11, 74, 315, Vijay, Vaibhav, Shweta, JGK-1, 2, JGG-1
☆	Pea	:	Rachna, Ambika, Subhra, Aparna, Paras, JP 855, KPMR 144-1, Vikash

- ☆ Lathyrus : Ratan, Pratik, Mahativda.
- ☆ Lentil : K 75, Lens 4076, Nuri, Sheri, JL-3.
- ☆ Soybean : Indira soya-9, JS-2, 335, 93-05, PK 472, Gourav, Ankur, Durga
- ☆ Groundnut : ICGS 1, 10, 11, 37, 44, SB-11, JL24, Chandra, Junagarh-11, Vikram, Verginia
- ☆ Sunflower : Modern, Jwalamukhi, MSFH-8, 17, KBSH-1, 44
- ☆ Safflower : JSF1, 2, 5, JS I7, JSH 129, Annagiri,
- ☆ Mustard : Pusa kalyani, Sufla, Kranti, Varuna, Krishna, Pusa bold, Vardan, Rohni
- ☆ Til : Selection-5, Krishna, JT-21, TC-25
- ☆ Ramtil : IGP-76, GP 10, JNS-1, 6
- ☆ Castor : Kranti, Jwala, Jyoti, JCH-4, DCH-32
- ☆ Linseed : RLC 92 (Indravati), Deepika, Kiran, Indira Alsi-32 (RLC-81), Jwahar 552
- ☆ Cotton : Anjali, Khandwa-2, Jwahar tapti, JKH-2, Pratima.
- ☆ Sugarcane : CO 671, COJ 64, 8338, 86-141, CO 86032, 62175

MUTANT VARIETIES OF CROPS

- ☆ Rice : *Jagannath, Prabhavati*
- ☆ Chickpea : BGM-48, BGM-413,
- ☆ Pea : Hans
- ☆ Arhar : Trombay, Vishakha-1
- ☆ Wheat : Sarbati sanora
- ☆ Cotton : MCU-7, MCU-10
- ☆ Tobacco : Jayshri, Bhavya
- ☆ Moong : Dhulli, Pant mung 2, MUM 2
- ☆ Urd : CO1, Sarla
- ☆ Mango : Rosica

IMPORTANT WEED FLORA OF CROPS

Sl.No.	Botanical Name	Family	English Name	Common Name
1.	*Achyranthes aspera*	Amaranthceae	Prickly chafflower	चिरचिटा/लटजीरा
2.	*Ageratum conyzoides*	Compositae	Bill goat weed	महकुआ/फुलनी
3.	*Amaranthus spinosus*	Amaranthceae	Spiny amaranthus	कांटेदार चौलाई

Sl.No.	Botanical Name	Family	English Name	Common Name
4.	*Amaranthus viridis*	Amaranthceae	Slender amaranthus	जंगली चौलाई
5.	*Agremone maxicana*	Papaveraceae	Mexican prickly poppy	सत्यानासी
6.	*Avena fatua*	Gramineae	Wild oat	जंगली जई
7.	*Boerhavia diffusa*	Nyctaginaceae	Spreading hog weed	विषखपरा
8.	*Brassica sinensis*	Cruciferae	Wild mustard	जंगली सरसों
9.	*Calotropis gigantean*	Ascletiabaceae	Giant swallow wort	आंक/मदार
10.	*Carthamus oxyacanthe*	Compositae	Wild safflower	जंगली कुसुम
11.	*Cassia tora*	Leguminosae	Buffalo gram	चरोटा/चकौड़ा
12.	*Celosia argentea*	Amaranthceae	White cock's comb	सिलयारी
13.	*Chenopodium album*	Chenopodiaceae	Lambsquate/Dog tooth grass	बथुआ
14.	*Cichorium intybus*	Compositae	Chicory/Blue daisy	कासनी
15.	*Convolvulus arvensis*	Convolvulaceae	Bind weed	हिरणखुरी
16.	*Corchorus acutangulus*	Tilliaceae	Wild jute	चेज
17.	*Cuscuta sp.*	Convolvulaceae	Dodder	अमरबेल
18.	*Cynodon dactylon*	Gramineae	Bermuda grass	दुब घास
19.	*Cyperyus rotundus*	Cyperaceae	Purple nutsedge	मोथा
20.	*Cyperus iria*	Cyperaceae	Yellow nutsedge/ Rice flat sedge	मोथा
21.	*Cyperyus difformis*	Cyperaceae	Umbrella sedge	मोथा
22.	*Datura alba*	Solanceae	Thorne apple	कांटेदार धतुरा
23.	*Datura stramonium*	Solanceae	Jimson weed	धतुरा
24.	*Dicanthum annulatum*	Gramineae	Marvel grass	कांदी
25.	*Digitaria sanguinalis*	Gramineae	Crab grass	घुड़-दुब
26.	*Echinochloa colonum*	Gramineae	Jungle rice	सांवा
27.	*Echinochloa crusgalli*	Gramineae	Barnyard grass	सांवा (मुंछवाला)

Sl.No.	Botanical Name	Family	English Name	Common Name
28.	*Eclipta alba*	Compositea	False daisy	भृंगराज/भंगड़ा
29.	*Eichhonia crassipes*	Pontederiaceae	Water hyacinth	जलकुंभी
30.	*Eleusine indica*	Gramineae	Goose grass	जंगली रागी
31.	*Euphorbia geniculata*	Compositae	Garden spurge	बड़ी दुधी
32.	*Euphorbia hirta*	Compositae	Pill pod spurge	छोटी दुधी
33.	*Ipomea repens*	Convolvulaceae	Swamp morning glory	जलकर्मी
34.	*Ischoemum rugosum*	Gramineae	Wrinkle grass	टोरा-टोरी
35.	*Lantana camara*	Verbenaceae	Prickly lantana	जरायन
36.	*Lathyrus sativus*	Leguminoceae	Lathyrus	खेसारी
37.	*Melilotus alba*	Leguminoceae	White sweet clover	सफेद सेंजी
38.	*Melilotus indica*	Leguminoceae	Yellow sweet clover	पीली सेंजी
39.	*Mimosa pudica*	Leguminoceae	Touch me not	लाजवंती
40.	*Mimosa spinosa*	Leguminoceae	Touch me not	लाजवंती (कांटेदार)
41.	*Ocimum camum*	Labiatae	Haory basin	बनतुलसी
42.	*Opuntia dilenaii*	Cacaceae	Prickly pear	नागफनी
43.	*Oryza sativa var fatua*	Gramineae	Wild rice	जंगली धान
44.	*Orobanche sp.*	Orobanchaceae	Broom rape	बिल्ली
45.	*Oxalis ocetorella*	Oxalidaceae	Sorrel	खट्टी-बुटी
46.	*Oxalis corniculata*	Oxalidaceae	Indian sorel	खट्टी-बुटी
47.	*Parthenium hysterophorus*	Compositae	Congress grass/ Wild carrot grass	गाजर घास
48.	*Paspalum sanguinale*	Gramineae	Knot grass	-
49.	*Phalaris minor*	Gramineae	Canary grass	गेहूं का मामा
50.	*Portulaca oleracea*	Portulaceae	Purslane	जंगली पालक
51.	*Portulaca quodrifolia*	Portulaceae	Purslane	नुनिया
52.	*Phyllanthus nururi*	Euphorbiaceae	Corn spurry/ Niruri	हजारदाना

Sl.No.	*Botanical Name*	*Family*	*English Name*	*Common Name*
53.	*Physalis minima*	Solanaceae	Ground cherry / Hog weed	चिरपोटी
54.	*Saccharum spontaneum*	Gramineae	Tiger grass	कांस
55.	*Sataria glauca*	Gramineae	Green fox tail	बंदरा-बंदरी
56.	*Sida rhombifolia*	Malvaceae	Sida	बरयारा (पीली)
57.	*Sida spinosa*	Malvaceae	Sida	बरयारा (हरा)
58.	*Solanum nigrum*	Solanaceae	Black night shade	मकोई
59.	*Solanum xanthocarpum*	Solanaceae	Prickly brinjal	भटकटैया
60.	*Sorghum halepanse*	Gramineae	Johnson grass	बरू
61.	*Spilanthus comelia*	Compositae	Wild mint	जंगली अकरकरा
62.	*Striga lutea*	Scrophulariaceae	Witch weed	अगिया
63.	*Trianthema monegyna*	—	Carpet weed	पथरचटा
64.	*Tridex procumbens*	Compositae	Mexican daisy	बारहमासी
65.	*Typha sp.*	Typhaceae	Cattail	टायफा
66.	*Vicia hirsute*	Leguminoceae	Common vetches	मुनमुना
67.	*Vicia sativa*	Leguminoceae	Vetches	टकरा अकरी
68.	*Xanthium strumerium*	Compositae	Cocklebur / Bur-weed	बड़ी गोखरू
69.	*Zizyphus rotundifolis*	Rhamnaceae	Wild ber	झरबेरी

IMPORTANT FOREST TREES

Sl.No.	*Botanical Name*	*Common Name*	*Toxicant Present*
1.	*Acacia catechu*	खैर	Tannin
2.	*Acacia leucophloea*	सफेद बबूल	Tannin
3.	*Acacia nilotica*	बबूल	Tannin
4.	*Aegle marmelos*	बेल	Tannin
5.	*Albizia lebbeck*	काला सिरस	Tannin
6.	*Albizia procera*	सफेद सिरस	Tannin

Sl.No.	Botanical Name	Common Name	Toxicant Present
7.	*Anthocephalus indicus*	कदम	—
8.	*Azadirachta indica*	नीम	Azadirachtin, Nimbin
9.	*Bambusa arundinacea*	बांस	HCN
10.	*Bauhinia variegata*	कचनार	Tannin
11.	*Butea monosperma*	पलास	—
12.	*Cassia fistula*	अमलतास	—
13.	*Delbergia sissoo*	शीशम	Tannin
14.	*Delonix regia*	गुलमैंहर	—
15.	*Emblica officinalis*	आंवला	—
16.	*Eucalyptus tereticornis*	नीलगीरी	—
17.	*Ficus banghalensis*	बरगद	Tannin
18.	*Ficus religiosa*	पीपल	Tannin
19.	*Leucaena leucocephala*	सूबबूल	Mimosine
20.	*Madhuca latifolia*	महुआ	Saponin
21.	*Mangifera indica*	आम	Amylase inhibitors
22.	*Morus alba*	मलबेरी	Tannin
23.	*Musa paradisiaca*	केला, बनाना	Amylase inhibitors, Seratonin
24.	*Polyalthia longifolia*	औंक	—
25.	*Pongamia pinnata*	करंज	Karanjin, Pongamol
26.	*Populus deltoides*	पोपुलर	—
27.	*Psidium guajava*	अमरूद	—
28.	*Pterocarpus marsupium*	बीजा	—
29.	*Shorea robusta*	साल	—
30.	*Syzygium cuminii*	जामुन	Tannin
31.	*Tamarandus indica*	इमली	—
32.	*Tectona grandis*	सागैंन (Teak)	—
33.	*Terminalia arjuna*	अर्जुन	—
34.	*Terminalia belirica*	बह ड़ा	—
35.	*Terminalia chaibula*	हर्रा	—

IMPORTANT MEDICINAL and AROMATIC CROPS

Sl.No.	Botanical Name	Common Name	Plant Part Used	Chemical Found
Medicinal crops				
1..	*Chlorophytum borivilianum*	Safed Musli	Root	Saponins
2.	*Withania somnifera*	Ashwagandha	Leaves & Root	Alkaloids
3.	*Rauvolfia serpentina*	Sarpagandha	Root (Dried)	Serpentine (alklo.)
4.	*Plantago ovata*	Isabgol	Husk of the seed	Glycoside
5.	*Acorus calamus*	Buch	Rhizome (Dried)	—
6.	*Bacopa morrieri*	Bramhi	Whole plant	Hydrolytin (alklo.)
7.	*Papaver somniferum*	Opium poppy	Latex & Seeds	Alkaloids
8.	*Strychnos nuxvomica*	Nux vomica	Seeds	Strychnine (alklo.)
Aromatic crops				
9.	*Cymbopogan flexuosus*	Lemon grass	Fresh grasses	Citral a & b
10.	*Mentha arvensis*	Pudina or Mint	Herbage	Menthol
11.	*Hibiscus sabadriffa*	Roselle	Fresh Calyces	Fatty oils
12.	*Vetiveria zizanoides*	Khus/ Vetiver	Root	Khusol, Vetiverone
13.	*Cymbopogan winterianus*	Citronella	Fresh herbage	Citronellol, Geraniol
14.	*Ferula foetida*	Asafoetida/ Hing	Gum resin	Organic sulpher
15.	*Ocimum sanctum*	Tulsi/ Basil	Leaves	Eugenol
16.	*Cymbopogon martini*	Palmarosa	Floral shoots	Geraniol

TEST WEIGHT OF CROPS

	Crop		Test Weight	Crop		Test Weight
☆	Rice	:	25	Linseed, Safflower	:	10
☆	Basmati rice	:	21	Lucerne	:	2-4
☆	Wheat, Barley, Oat	:	30-40	Sunflower	:	40-50
				Soybean	:	55

	Crop		Test Weight	Crop		Test Weight
☆	Cotton, Arhar	:	70-72	Moong	:	34-36
☆	Pea	:	100	Cowpea	:	80
☆	Mustard	:	3-5	Bajra	:	5-7
☆	Sorghum	:	25-30	French bean	:	38-44/1000 seeds
☆	Tobacco	:	2.5-3/10000 seeds			

☆ **Test weight**: weight of 1000 seeds of a crop

☆ **Seed Index**: weight of 100 seeds of a crop (used for bold seeded)

FRUIT TYPES AND EDIBLE PARTS OF DIFFERENT CROPS

	Crops/Fruits	*Fruit Types*	*Edible Parts*
☆	All cereals crops and grasses	Caryopsis	Endosperm and Embryo
☆	Most of leguminous crops *i.e.* gram, Pea, Arhar	Legume/ Pod	Seed/ Cotyledons
☆	Groundnut	Lomentum	Seed/ Cotyledons
☆	Mango	Drupe	Mesocarp
☆	Ber, Plum, Datpalm	Drupe	Apicarp and Mesocarp
☆	Tomato, Grape, Brinjal, Banana, Chilli	Berry	Pericarp and Placenta
☆	Papaya	Berry	Mesocarp
☆	Citrus spp.	Hesperidium	Endocarpic juicy hairs
☆	Pomegranate	Blusta	Aril
☆	Apple, Pear	Pome	Mansal thalamus
☆	Bael	Amphisarca	Succulent Placenta
☆	Cucurbits	Pipo	Apicarp and Mesocarp
☆	Coconut	Nut	Endosperm
☆	Jackfruit, Pineapple	Sorosis	Bracts/Perianth

ANTITRANSPIRENTS

1.	Stomatal closure type	:	2,4-D, Atrazine, PMA, Phosphon D, Potassium metabisulphate
2.	Film farming type	:	Hexadeconol, Cetyl alcohol, Paclobutrazole Mobileaf, Waxol, S-800, Hico-110R, Folicot, Silicon
3.	Reflectant type	:	Kaoline (5 per cent), China clay, Ca. bicarbonate, Lime water
4.	Growth retardant type	:	Cycocel (CCC), Phosphor

IMPORTANT METEOROLOGICAL INSTRUMENTS and THEIR USES

	Instruments		*Uses/ Measures*
☆	Altimeter	:	Height
☆	Aneroid barometer	:	Atmospheric pressure
☆	Anemometer	:	Wind speed/ velocity
☆	Auxanometer	:	Growth of plant
☆	Barograph	:	Continuous atmospheric pressure
☆	Crescograph	:	Growth of plant
☆	Cambel stokes recorder	:	Sunshine duration
☆	Drosometer	:	Dew
☆	Hygrometer/ Psychrometer	:	Relative humidity (RH)
☆	Evaporometer	:	Evapotranspiration
☆	Irrometer	:	Water stress, soil moisture tension
☆	Infiltrometer	:	Infiltration
☆	Lactometer	:	Fats % in milk
☆	Lysimeter	:	Evapotranspiration
☆	Pyrheliometer	:	Direct solar radiation
☆	PAR	:	Quantum sensor
☆	Psychrometer	:	Leaf water potential/ RH
☆	Porometer	:	Transpiration rate (ET)
☆	Pyranometer	:	Total incoming solar radiation
☆	Peizometer	:	Depth of water table

	Instruments		Uses/ Measures
☆	Rain gauge	:	Amount of rainfall
☆	Tensiometer	:	Soil moisture tension (0.8 bar)
☆	Wind vane	:	Wind direction
☆	Tensiometer	:	Soil moisture tension (0.8 bar)
☆	Wind vane	:	Wind direction

CLASSIFICATION OF SOIL PARTICLES
Based on size (mm)

Classification		IISS	USDA
Stone	:	> 250	> 250
Cobble	:	75–250	75–250
Gravel	:	2–75	2–75
Very course sand	:	–	1.0–2.0
Course sand	:	2–0.2	0.5–1.0
Fine sand	:	0.2–0.02	0.1–0.25
Very Fine sand	:	–	0.05–0.1
Silt	:	0.02–0.002	0.002–0.05
Clay	:	< 0.002	< 0.002

NUTRIENT CONTENTS OF COMMON FERTILIZERS

Sl.No.	Fertilizers		Nutrient Content (%)			
			N	P_2O_5	K_2O	S
Nitrogenous Fertilizers						
(A)	***Nitrate form***					
1.	Sodium nitrate	:	16.0	-	-	-
2.	Calcium nitrate	:	15.5	-	-	-
(B)	***Ammonical form***					
1.	Ammonium phosphate	:	16.0	20.0	-	-
2.	Ammonium chloride	:	24-26	-	-	-
3.	Ammonium sulphate	:	20.6	-	-	24
4.	Anhydrous ammonia	:	81.0	-	-	-

Sl.No.	Fertilizers		Nutrient Content (%)			
			N	P_2O_5	K_2O	S
(C)	**Ammonical Nitrate form**					
1.	Ammonium nitrate	:	33-34	-	-	-
2.	CAN	:	26.0	-	-	-
3.	Amm. sulphate nitrate	:	26.0	-	-	15
(D)	***Amide form***					
1.	Urea	:	46.0	-	-	-
2.	Calcium cyanide	:	21.0	-	-	-
Phosphatic Fertilizers						
(A)	***Water soluble***					
1.	SSP (single)	:	-	16-18	-	-
2.	DSP (double)	:	-	32.0	-	-
3.	TSP (triple)	:	-	46-48	-	-
4.	DAP	:	18	46	-	-
(B)	***Citric acid soluble***					
1.	Di calcium phosphate	:	-	34-39	-	-
2.	Basic slag	:	-	14-18	-	-
3.	Bone meal	:	-	23-30	-	-
(C)	***Insoluble***					
1.	Rock phosphate	:	-	20-40	-	-
2.	Rock bone meal	:	-	20-25	-	-
3.	Steamed bone meal	:	-	22.0	-	-
Phosphatic Fertilizers						
1.	Murate of potash/ KCl	:	-	-	60.0	-
2.	Sulphate of potash	:	-	-	48.0	-
3.	Potassium nitrate	:	-	-	44.0	-

CHEMICAL PROPERTIES OF SALINE, SODIC and ALKALINE SOILS

Types of Soil		*EC (dSm) at 25°C*	*ESP (%)*	*pH*
Saline soil	:	> 4	< 15	< 8.5
Saline alkaline soil/Sodic	:	> 4	> 15	< 8.5
Alkaline soil	:	< 4	> 15	8.5 – 10.0

FUNCTIONS OF PGRs

1.	Auxins	:	Cell division and root formation.
2.	Gibberellin	:	Cell division, breaking dormancy and cell elongation.
3.	Cytokinin	:	Delay senescence, breaking dormancy of seed and development of embryos in seed.
4.	Abscisic acid	:	Abscission of leaf and fruit, induce dormancy & maintain cell turgidity, facilitate stomata closure.
5.	Ethylene	:	Fruit ripening, iso-diametric growth of stems and roots.

DISEASES OR SYMPTOMS CAUSED DUE TO DEFICIENCY OF NUTRIENTS IN PLANTS

Deficiency of Nutrient		*Causes Diseases/Symptoms*
N	–	Buttoning in cauliflower
P	–	Sickle leaf disease.
K	–	Scorching and burning of leaves.
Ca	–	Blossom end rot in tomato and Ber, Tip hooking/burning in Cauliflower.
Mg	–	Sand drawn disease of Tobacco.
S	–	Tea yellow disease.
Fe	–	White eye of Paddy, Leaf bleaching in Sugarcane
Mn	–	Water core in Brassica, Marsh disease in Pea, Spotted yellow disease in Sugarbeet.
Cu	–	Dieback and little leaf in Citrus, Reclaimation disease in cereals.
Mo	–	Yellow spot disorder in Citrus, Whiptail disease in Cauliflower.
Zn	–	Little leaf in Brinjal and Mango, Bronzing in Guava, White bud in Maize, Khaira disease in Paddy.
Bo	–	Internal necrosis in Aonla and Mango, Browning in Cauliflower, Heat rot in Sugarbeet, Hen and Chicken disorder in Grape.

REVOLUTION IN AGRICULTURE

	Revolution		*Related to*
☆	Green revolution	:	Food grain production
☆	White revolution	:	Milk production
☆	Yellow revolution	:	Oilseeds production
☆	Gray revolution	:	Manures and Fertilizers
☆	Blue revolution	:	Fish production
☆	Red revolution	:	Meat/Tomato production
☆	Round revolution	:	Potato production
☆	Silver revolution	:	Egg production/Poultry
☆	Pink revolution	:	Prawn production
☆	Golden revolution	:	Fruit production (apple)
☆	Brown revolution	:	Non-conventional energy source
☆	Black revolution	:	Bio fuel (Jatropha) production
☆	Rainbow revolution	:	Agriculture (1996)
☆	Food chain revolution	:	Food grain production
☆	Evergreen revolution	:	Reduction in wastage of food grains, fruits and vegetables
☆	Parbhani revolution	:	Okra

IMPORTANT AGRICULTURAL DAYS

- ☆ Feb., 2 : National Wetland day
- ☆ Mar., 21 : World Forest day
- ☆ Apr., 22 : World Earth day
- ☆ Jun., 5 : World Environment day
- ☆ Jul., 1-7th : Van Mahotsava
- ☆ Sept., 16 : World Ozone day
- ☆ Oct., 16 : World Food day
- ☆ Dec., 23 : National Farmer's day
- ☆ Mar., 11 : Water Resources day
- ☆ Mar., 22 : World Water day
- ☆ May, 1 : International Labour day
- ☆ Jul., 1 : National Agricultural day
- ☆ Jul., 16 : ICAR day
- ☆ Oct. 4 : World Animal Welfare day
- ☆ Dec., 4 : Agriculture Women day

PRODUCTION OF FIELD CROPS IN INDIA (2010-11)

Sl.No.	Crops	Final Estimates		2nd Advance Estimates of
		Production (Mt.) 2008-09	Production (Mt.) 2009-10	Production (Mt.) 2010-11
1.	Rice	99.02	89.13	94.01
2.	Wheat	80.70	80.80	81.47
3.	Maize	19.70	16.70	-
4.	Sorghum	7.2	7.0	-
5.	Bajra	8.9	6.5	-
6.	Course cereals	39.48	33.77	40.08
7.	Arhar	2.3	2.6	-
8.	Gram	7.1	7.3	-
9.	Total pulses	14.60	14.66	16.51
10.	**Total foodgrains**	**234.40**	**218.20**	**232.07**
11.	Groundnut	7.2	5.5	-
12.	Rapseed & Mustard	7.2	6.4	-
13.	9 oilseeds (Total)	27.70	24.93	27.85
14.	Cotton	22.30	24.22	33.93
15.	Sugarcane	295.00	292.30	336.70
16.	Jute & Mesta	10.30	11.82	10.08

AREA, PRODUCTION & PRODUCTIVITY OF HORTICULTURAL CROPS IN INDIA

Sl.No.	Crops	2009-10 Final		
		Area (000 Mha)	Production (000 Mt.)	Productivity (Mt/ha)
Fruits				
1.	Mango	**2312**	15027	6.5
2.	Banana	770	**26470**	34.4
3.	Citrus	987	9638	9.8
4.	Guava	220	2572	11.4
5.	Grapes	106	881	8.3
6.	Litchi	74	483	6.5

Sl.No.	*Crops*	*2009-10 Final*		
		Area (000 Mha)	*Production (000 Mt.)*	*Productivity (Mt/ha)*
7.	Papaya	96	3913	**40.9**
8.	Pineapple	92	1387	15.1
9.	Pomegranate	125	820	6.6
10.	Sapota	159	1347	8.5
11.	Apple	283	1777	6.3
12.	Others	1105	7201	6.5
	Total	**6329**	**71516**	**11.3**
Vegetables				
1.	Potato	1835	36577	19.9
2.	Onion	756	12159	19.6
3.	Tomato	634	12433	16.6
4.	Brinjal	612	10563	17.2
5.	Cabbage	331	7281	22.0
6.	Cauliflower	348	6569	18.9
7.	Okra	452	4803	10.6
8.	Peas	365	3029	8.3
9.	Sweet Potato	119	1095	9.2
10.	Others	2300	31168	13.6
	Total	**7985**	**133738**	**16.7**
Aromatic:		509	573	1.1
Almond/Walnut:		142	193	1.4
Flowers Loose:		183	1021	-
Plantation Crops:		3265	11928	3.7
Spices:		2464	4016	1.6

% GROWTH OF HORTICULTURAL CROPS

Sl.No.	*Crops*	*09-10 over 08-09*	
		Area	*Production*
1.	Horticulture	1.0	3.9
2.	Fruit	3.7	4.5
3.	Vegetable	0.1	3.6

Availability of Agriculture Products/Capita/Day

Sl.No.	*Particular*	*Requirement*
1.	Cereals	407 g
2.	Pulses	37 g
3.	Fruits	120 g
4.	Vegetables	240 g
5.	Milk	263 g
6.	Egg	45 (no.)

Chapter 16
Recent Research/ Technologies in Agriculture

NANOTECHNOLOGY

- ☆ The term 'Nanotechnology' was coined by **Nario Taniguchi** in 1974 at Univ. of Tokyo.
- ☆ Nanotechnology is understanding and control of matter of dimension of 1-100 nm.
- ☆ Example of Nano based Smart Delivery System - **Halloysite**
- ☆ Nano Pesticide – Nano Particles (NPs) of ZnO, SiO_2 and TiO_2 used for Bacterial and Green algae.

BIO-INFORMATICS

- ☆ **Bio-informatics** is the application of computer science and information technology to the field of biology to the management of biological information.
- ☆ Computers are used to gather, store, analyze and integrate biological and genetic information which can then be applied to gene-based drug discovery and development.
- ☆ The primary goal of bioinformatics is to increase the understanding of biological processes and developing and applying computationally intensive techniques (*e.g.*, pattern recognition, data mining, machine learning algorithms, and visualization) to achieve this goal.

TRANSGENIC PLANTS/CROPS (GMO)

- ☆ **Transgenic plants** are crops which have been genetically modified with genes from another organism to make the plants more agriculturally productive.

- ☆ Transgenic plants are only those with genes from other species, whereas genetically modified plants can have both new genes and a re-arrangement of the genes already found in the plant.
- ☆ Transgenic plants have been developed for a variety of reasons: longer shelf life, disease resistance, herbicide resistance, pest resistance, and improved product quality.
- ☆ The First transgenic plant - **Flavr SavrTM tomato** for *delayed ripening* was released for commercial cultivation in 1994 by **Calgene** (Company).
- ☆ Crop having highest transgenic plant cultivation area – **Soybean** > Corn > Cotton

TERMINATOR TECHNOLOGY

- ☆ Terminator technology refers to research of seeds/plants that produce sterile seeds.
- ☆ This technology could be used to prevent any gene flow between biotechnology and traditional crops.
- ☆ Recently, it is used in **Cotton**.

HYDROPONICS

- ☆ **Hydroponics** is a method of growing plants using mineral nutrient solutions, in water, without soil.
- ☆ Terrestrial plants may be grown with their roots in the mineral nutrient solution only or in an inert medium, such as gravel, mineral wool or coconut husk.
- ☆ Hydroponics is a *subset of soil less culture.*

AEROPONICS

- ☆ **Aeroponics** is a system wherein roots are continuously or discontinuously kept in an environment saturated with fine drops (a mist or aerosol) of nutrient solution.
- ☆ The method requires no substrate and entails growing plants with their roots suspended in a deep air or growth chamber with the roots periodically wetted with a fine mist of atomized nutrients.
- ☆ Excellent aeration is the main advantage of aeroponics.

VERTICAL FARMING

- ☆ **Vertical farming** is a concept that argues that it is economically and environmentally viable to cultivate plant or animal life within skyscrapers, or on vertically inclined surfaces.

SYSTEM OF RICE INTENSIFICATION (SRI)

- ☆ The System of Rice Intensification is an alternative system for growing rice that produces substantially higher yields with **fewer plants** (planting far fewer seedlings per hill and per square meter) and with **fewer inputs** than either traditional methods, *ie.*, using less water, or more "modern" methods, requiring chemical fertilizer or agrochemicals.

- SRI is a combination of few practices that include changes in nursery management, seedling age while planting, planting method, spacing, water and nutrients management.
- The major components of SRI method are:
  - Planting of young seedlings (8-12 days old).
  - Planting single seedlings/hill along with soil carefully.
  - Wider spacing of 25cm×25cm.
  - Weeding with conoweeder to provide aeration and incorporation of biomass.
  - Applying mostly organic manures.
  - Water just at saturation point but no flooding.

AEROBIC RICE

- The main objectives are to improve the productivity and sustainability of rice-wheat cropping systems through increased efficiency of water and nutrient use.

  The aerobic rice practice includes:
  - Dry sowing of rice with minimum land preparation *i.e.* in non-puddled and non- flooded soil.
  - Efficient seed coating technology either with suitable Phosphobacterium and or Rhizobium cultures.
  - Square sowing with wider spacing to avoid root competition for crop growth.
  - Maintenance of moist soil but aerated soil during vegetative growth period.
  - Efficient weed management either by use of herbicide or by use of frequent hand weeding especially in the early stages of crop.
  - Allowing a thin film of water (1-2 cm) to be maintained after panicle initiation.

SUPER RICE

- *Super rice" is also k/s New Plant Type* (NPT).
- *"Super rice"* is a redesigned rice plant to break the yield –barriers of popular grown dwarf rice plant types and to face the new challenges of ever increasing population. In the 21st century. (acc. to Dr. G.S.KHUS)
- *Super* rice is a N.P.T. developed by IRRI that can produce yield of 12-15 tones/ha has 2-3 times greater no. of grains/panicle and thicker and sturdy stem.

The key aim of development of Super rice varieties is to increase per capita availability of rice and to a decline in real price of rice in International and Domestic markets.

SCUBA RICE

- Flooding affects 15–20 million hectares of lowland rice fields in Asia each year, it is a major contributor to the food insecurity and widespread poverty in these areas.
- IRRI scientists incorporated the *SUB1 gene* into popular local rice varieties collectively known as "scuba rice.
- Scuba rice varieties, which can survive up to 2 weeks of being under water, are now used by millions of farmers and serve as their first line of defence against flooding.

GOLDEN RICE

- Golden rice or GM rice is genetically engineered vitamin A rich rice.
- It was engineered to save million of children from blindness.

SUPER WHEAT

- 'Super varieties' of wheat resistant to the deadly stem rust fungus Ug99 and with up to 15 per cent better yields than today's varieties.

KISHAN KHAD

- It is also known as **CAN** (Calcium Ammonium Nitrate)
- Kishan khad is commercially prepared from ammonium nitrate and ground limestone or dolomite containing 20% nitrogen.
- It contains 26% nitrogen.
- One half of the nitrogen is in nitrate form and the remaining half in the ammonical form.
- It is almost neutral in nature.

Bt COTTON

- **Cotton with Bt gene** (*Bacillus thuringiensis*) is resistance against the pest, Helicoverpa.
- It is developed by U.S. based seed company 'Monsanto' and registered the name *'Bollgard'*.
- Bt variety obtained 25-27% more cotton along with reduced the cost of pesticides and protect environment from pesticidal hazards.

KISAN CALL CENTRE

- Kisan Call Centre (KCC), started since 21st **Jan 2004** (toll free No. 1551).

KISAN CREDIT CARD SCHEME (KCC)

- Kisan Credit Card Scheme was introduced **in August 1998**.
- KCC aims at providing adequate and timely support from the banking system to the farmers for their short-term credit needs for cultivation of crops.
- This mainly helps farmer for purchase of inputs etc., during the cropping season.
- Credit card scheme proposed to introduce flexibility to the system and improve cost efficiency.

NATIONAL AGRICULTURAL INSURANCE SCHEME

- National Agricultural Insurance Scheme was introduced in **1999-2000**.
- **Crop insurance** is purchased by agricultural producers, including farmers, ranchers, and others to protect themselves against either the loss of their crops due to natural disasters, such as hail, drought, and floods, or the loss of revenue due to declines in the prices of agricultural commodities.
- The two general categories of crop insurance are called crop-yield insurance and crop-revenue insurance.

PRECISION FARMING

- Precision farming means high tech agriculture, spatial variability management.
- It is the technique or method to find out the use of appropriate inputs, appropriate technology, decreasing cost of cultivation decisions, optimizing outputs for safety and security of food according to site or soil condition.

CONTRACT FARMING

- Contract farming is a system for the production and supply of agricultural products under forward contracts between cultivators and buyers.
- Here, the cultivator commits to provide an agricultural product of a specific type at a specific time and at a specified price that is required by the committed buyers.
- The main feature is that the contractor supplies all the material inputs and technical advice required for cultivation to the cultivator. In turn, the cultivator supplies the required land and labour.

LEISA

- LEISA stands for Low-External-Input Sustainable Agriculture.
- LEISA is an agricultural technique which makes optimal use of locally available natural and human resources (such as soil, water, vegetation, local plants and animals, and human labour, knowledge and skill) and

which is economically feasible, ecologically sound, culturally adapted and socially just.

GLOBAL WARMING

- *"Global warming is the extraordinary increase of Earth's surface temperature due to the increase of greenhouse gases concentration on the atmosphere."*
- **Greenhouse gases** (carbon dioxide, methane, nitrous oxide, and CFC) are the heat-trapping gases in the atmosphere.
- They trap the heat that came from the solar energy (sun radiation) which results continually warming of the earth.

OZONE DEPLETION

- Ozone layer is a protective layer in our atmosphere.
- It's about 19 to 30 km in distance from the Earth surface.
- It blocks the harmful ultraviolet (UV) rays that come from the sun.
- The concentration of the layer is usually under 10 parts ozone per million.
- Ozone layer concentration is measured by Ozonometer.
- The ozone layer is made up by the action of sunlight to oxygen, and the amount is stabled by the existence of nitrogen.
- If there was no ozone layer ever, cancer would dominate and even no life would be in this world.

ARTIFICIAL RAIN

- The clouds are injected with a seeding agent like dry ice, sodium chloride and silver iodide from an aircraft or using a ground generator for producing artificial rain.

ACID RAIN

- Acid rain (CO_2+*Rain drops*) basically have Carbonic acid with pH of less than 5.6.
- This is caused by the presence of air pollutants, like sulphur dioxide and nitrogen oxides. They produce acids if combined with water.